# Math Mammoth
# Grade 8-B Worktext

## Canadian Version

*By Maria Miller*

Photo credit:

p. 115, map of Nashville © OpenStreetMap contributors
Licensed under the license at www.openstreetmap.org/copyright

# Contents

# Chapter 7: Systems of Linear Equations

# Chapter 8: Bivariate Data

# Foreword

Math Mammoth Grade 8 comprises a complete math curriculum for the eighth grade mathematics studies. The curriculum meets the Common Core standards.

In 8th grade, students spend the majority of the time with algebraic topics, such as linear equations, functions, and systems of equations. The other major topics are geometry and statistics.

The main areas of study in Math Mammoth Grade 8 are:

- Exponents laws and scientific notation
- Square roots, cube roots, and irrational numbers
- Geometry: congruent transformations, dilations, angle relationships, volume of certain solids, and the Pythagorean Theorem
- Solving and graphing linear equations;
- Introduction to functions;
- Systems of linear equations;
- Bivariate data.

This book, 8-B, covers the topic of graphing linear equations. The focus is on the concept of slope.

In chapter 6, our focus is on square roots, cube roots, the concept of irrational numbers, and the Pythagorean Theorem and its applications.

Next, in chapter 7, students solve systems of linear equations, using both graphing and algebraic techniques. There are also lots of word problems that are solved using a pair of linear equations.

The last chapter then delves into bivariate data. First, we study scatter plots, which are based on numerical data of two variables. Then we look at two-way tables, which are built from categorical bivariate data.

Part 8-A covers exponent laws, scientific notation, geometry, linear equations, and an introduction to functions.

I heartily recommend that you read the full user guide in the following pages.

*I wish you success in teaching math!*

*Maria Miller, the author*

# User Guide

Note: You can also find the information that follows online, at https://www.mathmammoth.com/userguides/ .

## Basic principles in using Math Mammoth Complete Curriculum

Math Mammoth is mastery-based, which means it concentrates on a few major topics at a time, in order to study them in depth. The two books (parts A and B) are like a "framework", but you still have some liberty in planning your student's studies. In eighth grade, chapters 2 (geometry), 3 (linear equations) and chapter 4 (functions) should be studied before chapter 5 (graphing linear equations). Also, chapters 3, 4, and 5 should be studied before chapter 7 (systems of linear equations) and before chapter 8 (statistics). However, you still have some flexibility in scheduling the various chapters.

Math Mammoth is not a scripted curriculum. In other words, it is not spelling out in exact detail what the teacher is to do or say. Instead, Math Mammoth gives you, the teacher, various tools for teaching:

- **The two student worktexts** (parts A and B) contain all the lesson material and exercises. They include the explanations of the concepts (the teaching part) in blue boxes. The worktexts also contain some advice for the teacher in the "Introduction" of each chapter.

  The teacher can read the teaching part of each lesson before the lesson, or read and study it together with the student in the lesson, or let the student read and study on his own. If you are a classroom teacher, you can copy the examples from the "blue teaching boxes" to the board and go through them on the board.

- Don't automatically assign all the exercises. Use your judgement, trying to assign just enough for your student's needs. You can use the skipped exercises later for review. For most students, I recommend to start out by assigning about half of the available exercises. Adjust as necessary.

- For each chapter, there is a **link list to various free online games** and activities. These games can be used to supplement the math lessons, for learning math facts, or just for some fun. Each chapter introduction (in the student worktext) contains a link to the list corresponding to that chapter.

- The student books contain some **mixed review lessons**, and the curriculum also provides you with additional **cumulative review lessons**.

- There is a **chapter test** for each chapter of the curriculum, and a comprehensive end-of-year test.

- You can use the free online exercises at https://www.mathmammoth.com/practice/
  This is an expanding section of the site, so check often to see what new topics we are adding to it!

- There are answer keys for everything.

## How to get started

Have ready the first lesson from the student worktext. Go over the first teaching part (within the blue boxes) together with your student. Go through a few of the first exercises together, and then assign some problems for the student to do on their own.

Repeat this if the lesson has other blue teaching boxes.

Many students can eventually study the lessons completely on their own — the curriculum becomes self-teaching. However, students definitely vary in how much they need someone to be there to actually teach them.

## Pacing the curriculum

Each chapter introduction contains a suggested pacing guide for that chapter. You will see a summary on the right. (This summary does not include time for optional tests.)

Most lessons are 3 or 4 pages long, intended for one day. Some lessons are 5 pages and can be covered in two days.

It can also be helpful to calculate a general guideline as to how many pages per week the student should cover in order to go through the curriculum in one school year.

| Worktext 8-A | | Worktext 8-B | |
|---|---|---|---|
| Chapter 1 | 13 days | Chapter 5 | 15 days |
| Chapter 2 | 27 days | Chapter 6 | 16 days |
| Chapter 3 | 21 days | Chapter 7 | 17 days |
| Chapter 4 | 14 days | Chapter 8 | 11 days |
| TOTAL | 75 days | TOTAL | 59 days |

The table below lists how many pages there are for the student to finish in this particular grade level, and gives you a guideline for how many pages per day to finish, assuming a 160-day (32-week) school year. The page count in the table below *includes* the optional lessons.

**Example:**

| Grade level | School days | Days for tests and reviews | Lesson pages | Days for the student book | Pages to study per day | Pages to study per week |
|---|---|---|---|---|---|---|
| 8-A | 84 | 8 | 214 | 76 | 2.8 | 14.1 |
| 8-B | 76 | 8 | 189 | 68 | 2.8 | 13.9 |
| Grade 8 total | 160 | 16 | 403 | 144 | 2.8 | 14 |

The table below is for you to fill in. Allow several days for tests and additional review before tests — I suggest at least twice the number of chapters in the curriculum. Then, to get a count of "pages to study per day", **divide the number of lesson pages by the number of days for the student book**. Lastly, multiply this number by 5 to get the approximate page count to cover in a week.

| Grade level | Number of school days | Days for tests and reviews | Lesson pages | Days for the student book | Pages to study per day | Pages to study per week |
|---|---|---|---|---|---|---|
| 8-A | | | 214 | | | |
| 8-B | | | 189 | | | |
| Grade 8 total | | | 403 | | | |

Now, something important. Whenever the curriculum has lots of similar practice problems (a large set of problems), feel free to **only assign 1/2 or 2/3 of those problems**. If your student gets it with less amount of exercises, then that is perfect! If not, you can always assign the rest of the problems for some other day. In fact, you could even use these unassigned problems the next week or next month for some additional review.

In general, 8th graders might spend 45-75 minutes a day on math. If your student finds math enjoyable, they can of course spend more time with it! However, it is not good to drag out the lessons on a regular basis, because that can then affect the student's attitude towards math.

## Using tests

For each chapter, there is a **chapter test**, which can be administered right after studying the chapter. **The tests are optional.** The main reason for the tests is for diagnostic purposes, and for record keeping. These tests are not aligned or matched to any standards.

In the digital version of the curriculum, the tests are provided both as PDF files and as html files. Normally, you would use the PDF files. The html files are included so you can edit them (in a word processor such as Word or LibreOffice), in case you want your student to take the test a second time. Remember to save the edited file under a different file name, or you will lose the original.

The end-of-year test is best administered as a diagnostic or assessment test, which will tell you how well the student remembers and has mastered the mathematics content of the entire grade level.

## Using cumulative reviews and the worksheet maker

The student books contain mixed review lessons which review concepts from earlier chapters. The curriculum also comes with additional cumulative review lessons, which are just like the mixed review lessons in the student books, with a mix of problems covering various topics. These are found in their own folder in the digital version, and in the Tests & Cumulative Reviews book in the printed version.

The cumulative reviews are optional; use them as needed. They are named indicating which chapters of the main curriculum the problems in the review come from. For example, "Cumulative Review, Chapter 4" includes problems that cover topics from chapters 1-4.

Both the mixed and cumulative reviews allow you to spot areas that the student has not grasped well or has forgotten. When you find such a topic or concept, you have several options:

1. Check for any online games and resources in the Introduction part of the particular chapter in which this topic or concept was taught.

2. If you have the digital version, you could simply reprint the lesson from the student worktext, and have the student restudy that.

3. Perhaps you only assigned 1/2 or 2/3 of the exercise sets in the student book at first, and can now use the remaining exercises.

4. Check if our online practice area at https://www.mathmammoth.com/practice/ has something for that topic.

5. Khan Academy has free online exercises, articles, and videos for most any math topic imaginable.

## Concerning challenging word problems and puzzles

While this is not absolutely necessary, I heartily recommend supplementing Math Mammoth with challenging word problems and puzzles. You could do that once a month, for example, or more often if the student enjoys it.

The goal of challenging story problems and puzzles is to **develop the student's logical and abstract thinking and mental discipline**. I recommend starting these in fourth grade, at the latest. Then, students are able to read the problems on their own and have developed mathematical knowledge in many different areas. Of course I am not discouraging students from doing such in earlier grades, either.

Math Mammoth curriculum contains lots of word problems, and they are usually multi-step problems. Several of the lessons utilize a bar model for solving problems. Even so, the problems I have created are usually tied to a specific concept or concepts. I feel students can benefit from solving problems and puzzles that require them to think "outside of the box" or are just different from the ones I have written.

I recommend you use the free Math Stars problem-solving newsletters as one of the main resources for puzzles and challenging problems:

**Math Stars Problem Solving Newsletter (grades 1-8)**
https://www.homeschoolmath.net/teaching/math-stars.php

I have also compiled a list of other resources for problem solving practice, which you can access at this link:

https://l.mathmammoth.com/challengingproblems

Another idea: you can find puzzles online by searching for "brain puzzles for kids," "logic puzzles for kids" or "brain teasers for kids."

## Frequently asked questions and contacting us

If you have more questions, please first check the FAQ at https://www.mathmammoth.com/faq-lightblue

If the FAQ does not cover your question, you can then contact us using the contact form at the Math Mammoth.com website.

# Chapter 5: Graphing Linear Equations
## Introduction

This chapter focuses on how to graph linear equations, and in particular, on the concept of slope in that context.

We start by graphing and comparing proportional relationships, which have the equation of the form $y = mx$. Students are already familiar with these, and know that $m$ is the constant of proportionality. In this chapter, they learn that $m$ is also the slope of the line, which is a measure of its steepness.

Then we go on to study slope in detail, its definition as the ratio of the change in $y$-values and the change in $x$-values. Students learn that it doesn't matter which two points on a line you use to calculate the slope, and study a geometric proof of this fact. They practise drawing a line with a given slope and that goes through a given point, and determine if three given points fall on the same line.

Then it is time to study the slope-intercept equation of a line, and connect the idea of an initial value of a function (chapter 4) with the concept of $y$-intercept in the context of graphing. Students graph lines given in the slope-intercept form, and write equations of lines from their graphs.

Next, we study horizontal and vertical lines and their simple equations. The standard form of a linear equation follows next. The last major topic is how the slope reveals to us whether two lines are parallel or perpendicular to each other.

## Pacing Suggestion for Chapter 5

This table does not include the chapter test as it is found in a different book (or file).
Please add one day to the pacing if you use the test.

| The Lessons in Chapter 5 | page | span | suggested pacing | your pacing |
|---|---|---|---|---|
| Graphing Proportional Relationships 1 | 13 | *3 pages* | 1 day | |
| Graphing Proportional Relationships 2 | 16 | *3 pages* | 1 day | |
| Comparing Proportional Relationships | 19 | *4 pages* | 1 day | |
| Slope, Part 1 | 23 | *4 pages* | 1 day | |
| Slope, Part 2 | 27 | *3 pages* | 1 day | |
| Slope, Part 3 | 30 | *5 pages* | 2 days | |
| Slope-Intercept Equation 1 | 35 | *4 pages* | 1 day | |
| Slope-Intercept Equation 2 | 39 | *3 pages* | 1 day | |
| Write the Slope-Intercept Equation | 42 | *3 pages* | 1 day | |
| Horizontal and Vertical Lines | 45 | *3 pages* | 1 day | |
| The Standard Form | 48 | *3 pages* | 1 day | |
| More Practice (optional) | 51 | *(2 pages)* | (1 day) | |
| Parallel and Perpendicular Lines | 53 | *3 pages* | 1 day | |
| Mixed Review Chapter 5 | 56 | *3 pages* | 1 day | |
| Chapter 5 Review | 59 | *4 pages* | 1 day | |
| Chapter 5 Test (optional) | | | | |
| **TOTALS** | | *48 pages* | 15 days | |
| *with optional content* | | *(50 pages)* | (16 days) | |

## Helpful Resources on the Internet

We have compiled a list of Internet resources that match the topics in this chapter, including pages that offer:

- **online practice** for concepts;
- online **games**, or occasionally, printable games;
- **animations** and interactive **illustrations** of math concepts;
- **articles** that teach a math concept.

We heartily recommend you take a look! Many of our customers love using these resources to supplement the bookwork. You can use these resources as you see fit for extra practice, to illustrate a concept better and even just for some fun. Enjoy!

https://l.mathmammoth.com/gr8ch5

Scan me

# Graphing Proportional Relationships 1

We will now review what it means when two variables are **in direct variation** or **in proportion**. The basic idea is that whenever one variable changes, the other varies (changes) proportionally or at the same rate.

**Example 1.** The wholesaler posted the following table for the price of potatoes:

| weight (kg) | 5 | 10 | 15 | 20 | 25 | 30 |
|---|---|---|---|---|---|---|
| cost | $5.50 | $11.00 | $16.50 | $22.00 | $27.50 | $33.00 |

Each pair of cost and weight forms a rate — and so does each pair of weight and cost. However, it is more common to look at the rate "cost over weight", such as $27.50/(25 kg), than vice versa.

If all of the rates in the table are equivalent, then the weight and the cost *are* proportional.

To check for that, we have several means. One is to calculate **the unit rate** (the rate for 1 kg) from each of these rates, and check whether you get the same unit rate.

In this case, that is so. The unit rate is $1.10/kg, no matter which rate from the table we'd use to calculate it.

One other way to check is, if one quantity doubles (or triples), will the other double (or triple) also? This is especially useful for noticing if the quantities are *not* in direct variation.

**Example 2.** Here, when the weight doubles from 5 kg to 10 kg, the price also doubles. But what happens with the price when the weight doubles from 10 kg to 20 kg?

| weight (kg) | 5 | 10 | 15 | 20 | 25 | 30 |
|---|---|---|---|---|---|---|
| cost | $6 | $12 | $18 | $22 | $26 | $30 |

The price does not double! So, the quantities are not in proportion.
The seller is giving you some discount if you purchase higher quantities.

Also, if you calculate the unit rate from $6/(5 kg) and from $22/(20 kg), they are not equal. (Verify this.)

1. Are the quantities in a proportional relationship? If yes, list the unit rate.

a.

| time (hr) | 0 | 1 | 2 | 3 | 4 | 5 |
|---|---|---|---|---|---|---|
| distance (km) | 0 | 50 | 90 | 140 | 190 | 240 |

b.

| time (hr) | 0 | 1 | 2 | 3 | 4 | 5 |
|---|---|---|---|---|---|---|
| distance (km) | 0 | 45 | 90 | 135 | 180 | 225 |

c.

| age (days) | 0 | 1 | 2 | 3 | 4 | 5 | 6 | 7 |
|---|---|---|---|---|---|---|---|---|
| height (cm) | 0 | 0 | 0 | 2 | 4 | 6 | 8 | 10 |

d.

| length (m) | 0 | 0.5 | 1 | 1.5 | 2 | 4 | 5 | 10 |
|---|---|---|---|---|---|---|---|---|
| cost ($) | 0 | 3 | 6 | 9 | 12 | 24 | 30 | 60 |

2. Now consider the tables of values in #1 as functions, where the variable listed on top is the independent variable. For the ones where the quantities were in proportion, calculate the rate of change.

   What is its relationship to the unit rate?

When two quantities are in a proportional relationship, or in direct variation (the two terms are synonymous):

**(1) Each rate formed by the quantities is equivalent to any other rate of the quantities.**

**(2) The equation relating the two quantities is of the form $y = mx$, where $y$ and $x$ are the variables, and $m$ is a constant. The constant $m$ is called the constant of proportionality and is also the unit rate.**

**(3) When plotted, the graph is a straight line that goes through the origin.**

3. Choose an equation from below where the variables $x$ and $y$ are in direct variation (proportional):

$$y = \frac{3}{x} \qquad\qquad y = 3x$$

$$xy = 3 \qquad\qquad y = x^3$$

Then graph that equation in the grid.

*Hint:* The point $(0, 0)$ is always included in direct variation. All you need to do is plot one other point, and then draw a line through the origin and that point.

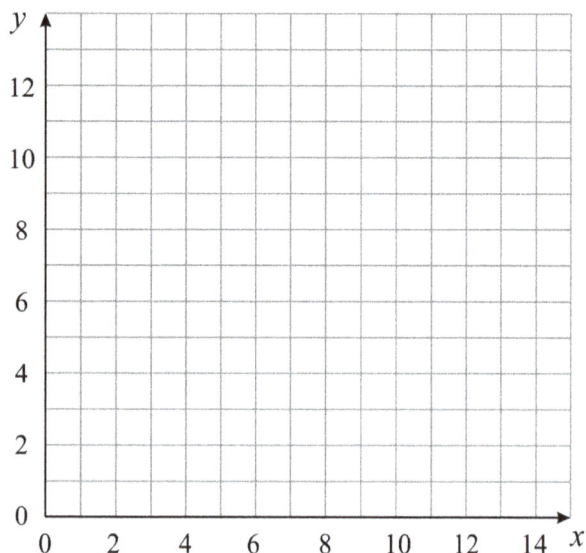

4. Choose the representations that show a proportional relationship.

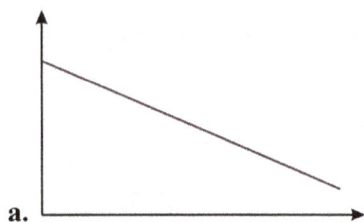

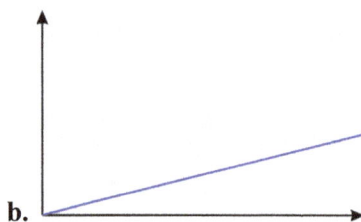

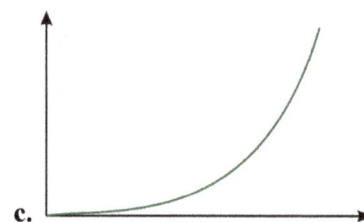

a.  b.  c.

d.

| $x$ | 0 | 1 | 2 | 3 | 4 | 5 |
|---|---|---|---|---|---|---|
| $y$ | 15 | 17 | 19 | 21 | 23 | 25 |

e. $y = 2x + 9$

f. $y = (3/4)x$

g.

| $x$ | 0 | 4 | 8 | 12 | 16 | 20 |
|---|---|---|---|---|---|---|
| $y$ | 0 | 3 | 6 | 9 | 12 | 15 |

5. Two of the above representations are the exact same relationship. Which ones?

**Example 2.** In a direct variation, $y = 9$ when $x = 12$. Write an equation for the relationship.

Since this is direct variation (proportional relationship), the equation is of the form $y = mx$, where $m$ is the constant of proportionality.

The constant of proportionality is the ratio **(dependent variable)/(independent variable)**, so in this case it is $y/x = 9/12$, or $3/4$. So, the equation is $y = (3/4)x$.

At this point, it is good to check that the point $(12, 9)$ satisfies the equation, to check for errors: Is it true that $9 = (3/4) \cdot 12$ ? Yes, it is.

To graph the equation, we could simply plot the point $(12, 9)$, and draw a line through it and the origin.

6. In a direct variation, when $x$ is 14, $y$ is 10.

   **a.** Write an equation for this proportional relationship.

   **b.** Graph a line for this relationship in the grid.

   **c.** What is $x$ when $y = 40$?

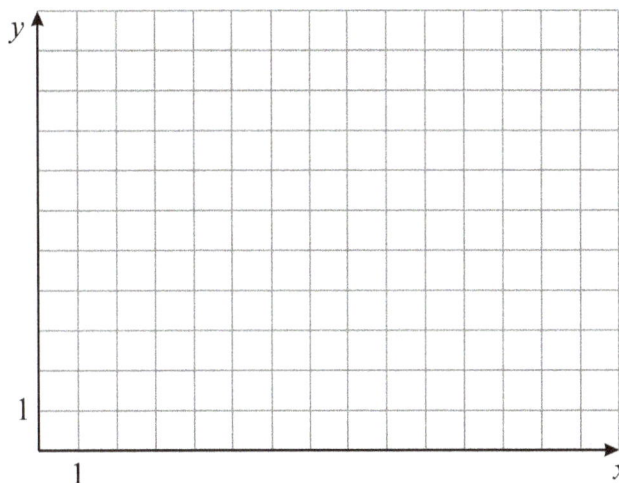

7. Organic rolled oats cost $34 for 4 kg.

   **a.** Write an equation for this proportional relationship, using the variables C for cost and $w$ for the amount (weight) of oats.

   **b.** Graph the equation in the grid. Design the scaling on the cost-axis so that the point corresponding to 12 kg fits on the grid.

   **c.** How much do 15 kg of the oats cost?

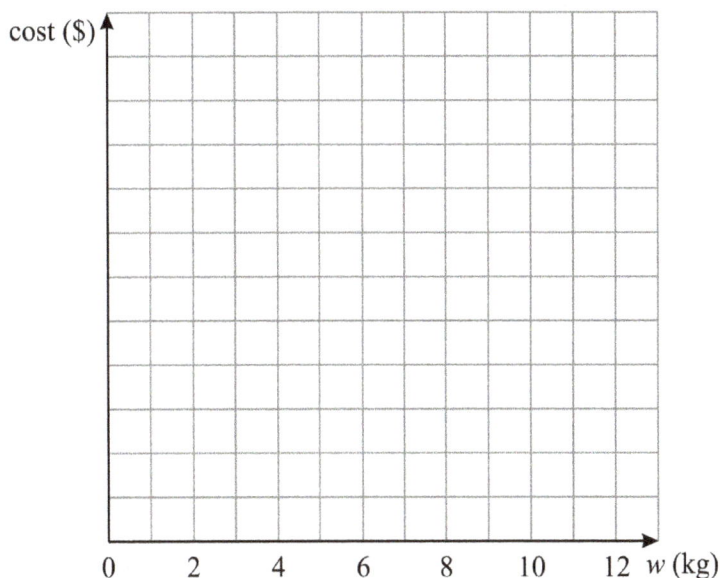

8. If $y$ is 120 when $x$ is 400 in a direct variation, then what is $y$ when $x$ is 80?

# Graphing Proportional Relationships 2

**Example 1.** The graph for the cost of apples as a function of their weight is a line through the origin, which means the cost and the weight are proportional.

The equation for the graph is C = 3.5$w$. So, the constant of proportionality is 3.5, which is also the rate of change of this function.

The unit rate is $3.50/kg. This corresponds to the point (1, 3.5) on the graph.

We also talk about **the slope of the line**, which is a measure of the steepness of the line, or how quickly it rises upwards (or slopes downwards).

It is the same idea as the rate of change, but in the context of graphing. Here, for each 1-kg increase in weight, the cost increases by $3.50. This means the slope of this line is 3.5 — the same as the rate of change, and the unit rate.

We will look at slope in more detail in other lessons.

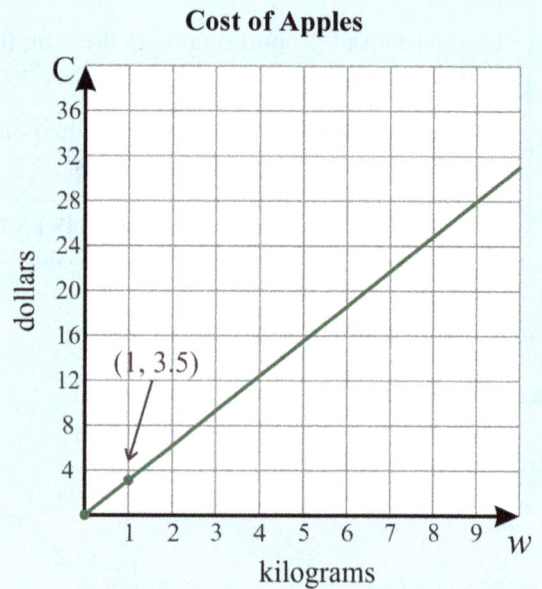

**Cost of Apples**

1. The graph shows the distance a caterpillar has crawled over time.

   **a.** Are the quantities *time* and *distance* proportional?

      How can you tell?

   **b.** What is the caterpillar's speed? Use the same units as in the graph.

      What is the unit rate?

      What is the slope of the line?

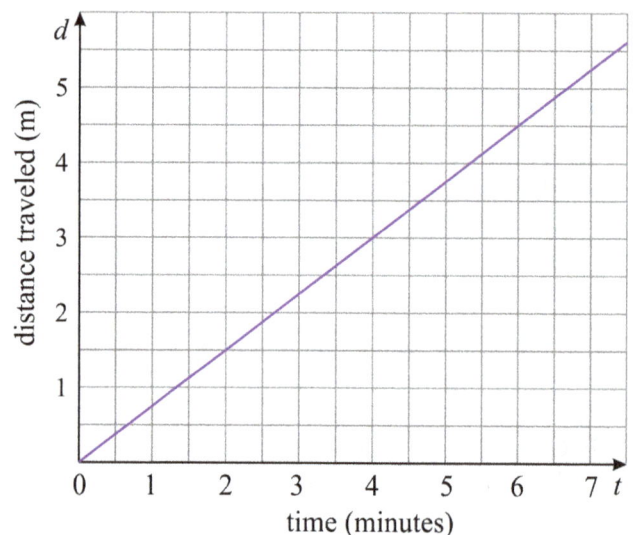

   **c.** Write an equation for the graph.

   **d.** Continuing with the same speed, how long will the caterpillar take to travel 9.5 metres?

2. Another caterpillar crawls 4.5 metres in 5 minutes. Draw another line in the grid for question #2, for the distance that this second caterpillar crawls over time, going with the same speed. Is this second caterpillar faster than the first?

3. A fabric costs $5.50 per metre.

    **a.** Write an equation for the cost as a function of the length of the fabric.

    **b.** Plot your equation. Design the scaling of the axes so that the point for 5 metres of fabric fits on the graph.

    **c.** What is the slope in this situation?

    What is the rate of change?

    **d.** Plot the point corresponding to the unit rate.

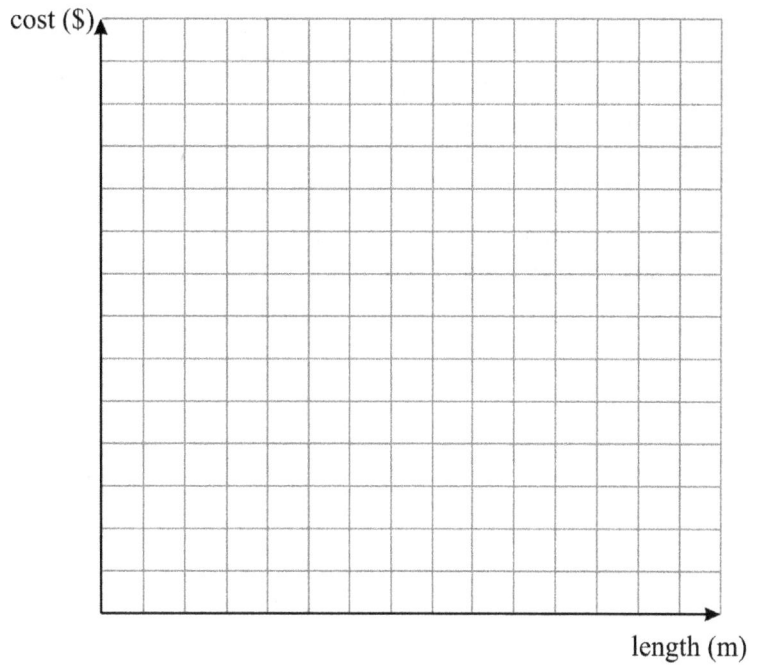

cost ($)

length (m)

4. Juan rode his motorcycle with a constant speed. After 12 minutes, he had travelled 9 km.

    **a.** Is the relationship between distance and time proportional?

    **b.** What is his speed? Include the units.

    **c.** What is the unit rate and the slope?

    **d.** Write an equation for the distance he has covered as a function of time.

    **e.** Graph your equation. Make it so that the point corresponding to 30 minutes fits on the graph.

    **f.** How long will it take him to travel 24 km?

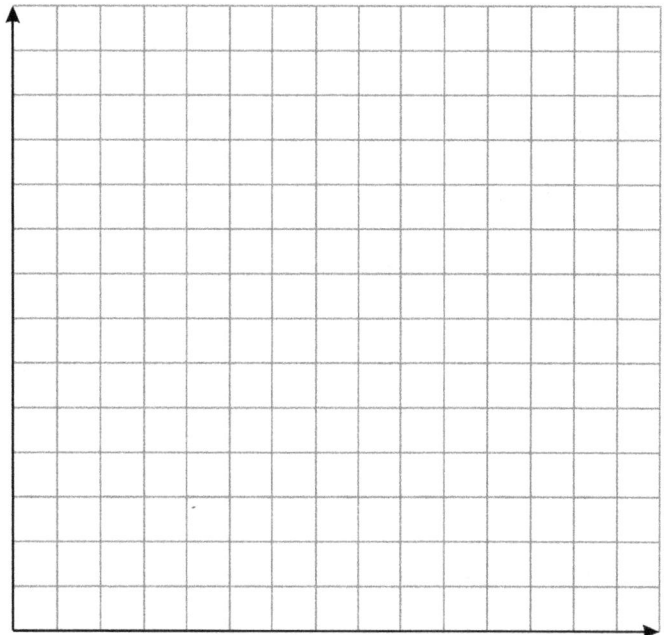

17

5. The Garcia family used 1500 ml of shampoo in six months.

   **a.** Considering the amount of shampoo used over time, what is the unit rate? Use the same units as in the graph.

   **b.** Write an equation for the amount of shampoo as a function of time.

   **c.** Plot your graph.

   **d.** What is the slope?

   **e.** How long will it take for the family to use 5 litres of shampoo?

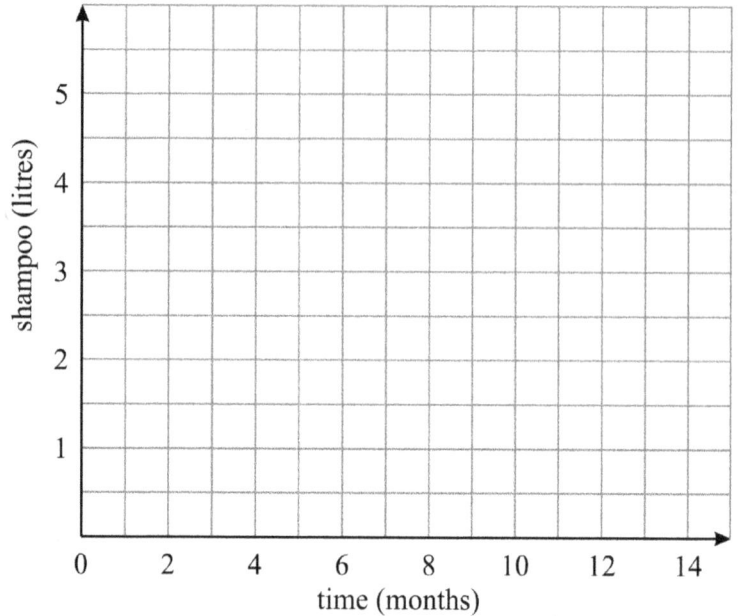

6. Elaine used half a litre of concentrate and 2 litres of water to make 2.5 litres of orange juice.

   There are three quantities here — the amount of concentrate, the amount of water, and the amount of juice — but let's only consider the first and the last.

   **a.** Plot a graph showing the relationship between the amount of concentrate (C) and the amount of juice (J).

   **b.** Write an equation for your graph.

   **c.** What is the unit rate?

   **d.** How much concentrate and how much water will Elaine need to make 6 litres of juice?

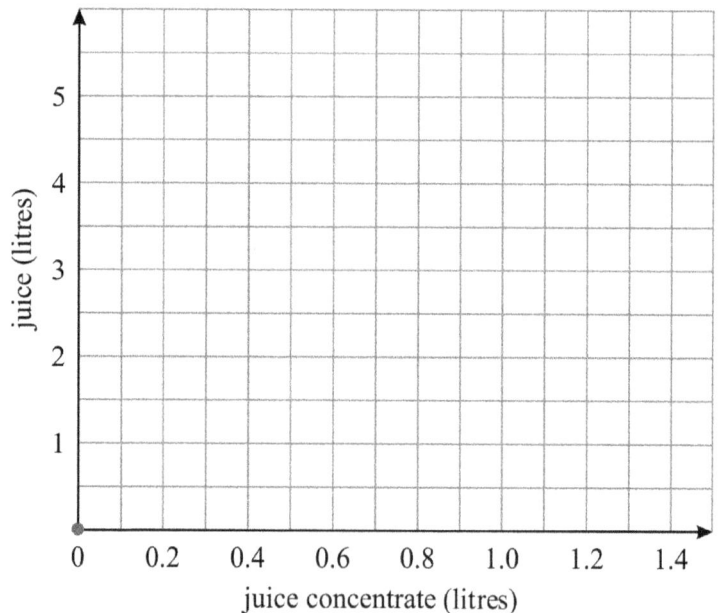

# Comparing Proportional Relationships

1. The table shows the cost of watermelon as a function of its weight. The price of bananas is given by the equation P = 1.6w where w is the weight in kilograms.

**Price of watermelon**

| weight (kg) | 0 | 2 | 4 | 6 | 8 |
|---|---|---|---|---|---|
| cost ($) | 0 | 3.60 | 7.20 | 10.80 | 14.40 |

**a.** Which fruit costs more for 5 kg of it? How much more?

**b.** Find the unit price for both.

2. Andrew is considering two different kinds of paint for the rooms of his house. The table shows how much wall area Paint 1 covers. The graph shows how much Paint 2 covers.

**Paint 1**

| area (m²) | paint (L) |
|---|---|
| 17.5 | 2 |
| 35 | 4 |
| 52.5 | 6 |
| 70 | 8 |
| 105 | 12 |
| 140 | 16 |
| 175 | 20 |
| 210 | 24 |
| 350 | 40 |

**Paint 2**

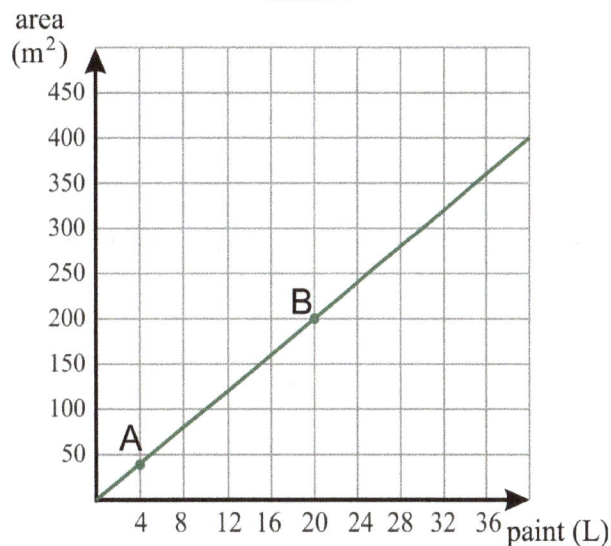

**a.** Which paint covers more wall area with 36 litres of paint?

**b.** The walls of a bedroom are 32 m². Calculate how much of each kind of paint would be needed to paint the room.

**c.** Draw a line depicting the area that Paint 1 covers, in the same grid as for Paint 2.

**d.** Estimate using the graphs about how much more the one paint covers than the other, with 32 litres.

3. Let's compare the speeds at which Henry and Jerry are travelling. Henry travels on his tractor at a constant speed of 25 km/h. The graph shows you the distance Jerry has covered over time on his moped.

   a. Which vehicle travels faster?

      How much faster?

   b. Graph a line depicting the distance Henry has travelled over time, in the same grid.

   c. State the slope of both lines.

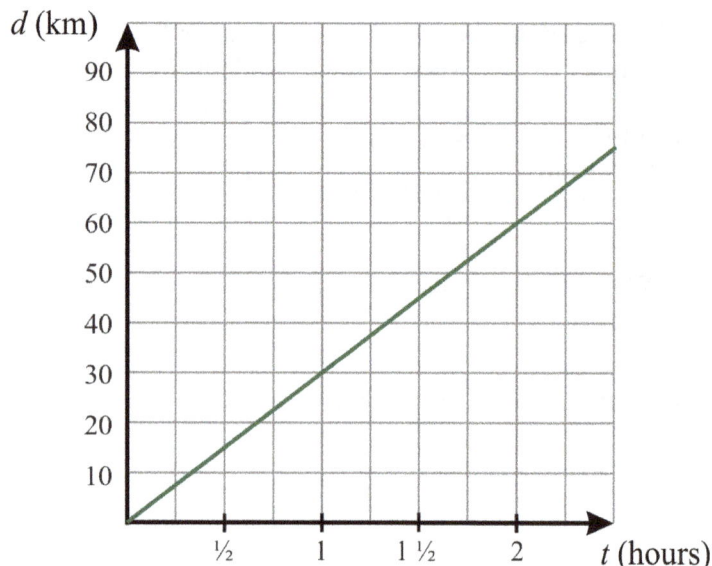

   d. Between these two vehicles, how much faster will the faster vehicle travel a distance of 10 km?

4. Jeanine rides a bike from her home to her grandma's, who lives 14 km away. The graph shows the distance she has covered over time. Her brother William goes the same distance on a motorcycle, going at a constant speed of 18 km per half an hour.

   a. Who travels faster?

   b. Plot a graph of the distance William covers over time in this grid.

   c. State the slope of both lines.

   d. How much quicker will William get to their grandma's place than Jeanine?

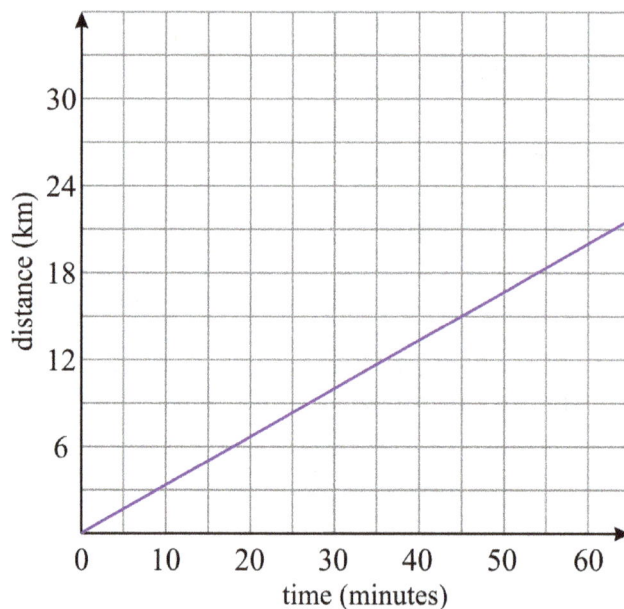

5. Blueberries cost $12.00/kg and strawberries cost $8/kg.

  **a.** Draw a graph for the cost of blueberries as a function of their weight, and another for the strawberries. Be sure to design the scaling for the cost axis to fit the points corresponding to 6 kg of either kind of berry on the grid.

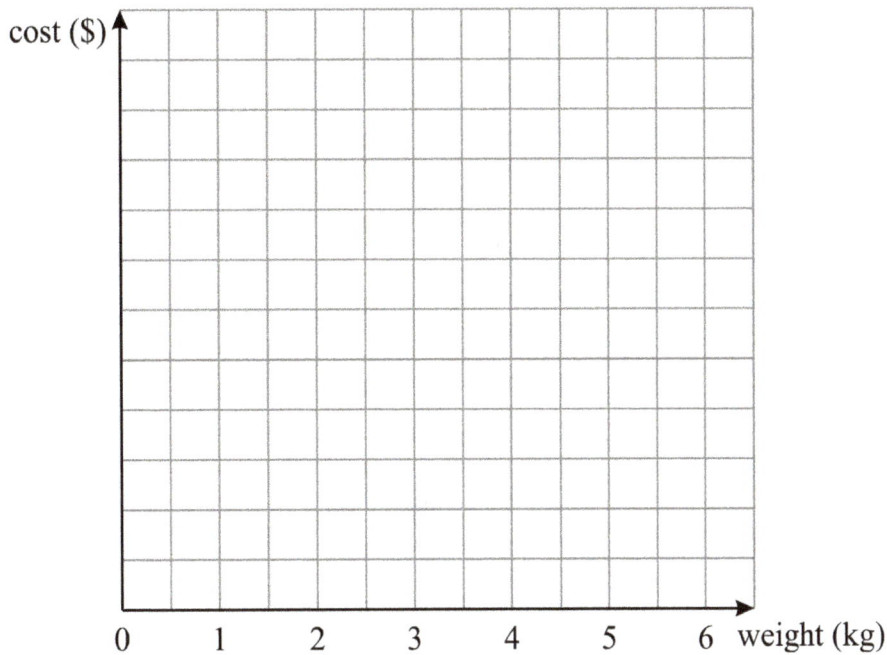

  **b.** Write an equation relating the cost and the weight, separately for each kind of berry.

  **c.** State the slopes of the two lines.

  **d.** How can you see in the graph which berries cost more?

  **e.** Use the graph to estimate the cost difference between 4 kg of strawberries versus 4 kg of blueberries.

6. **a.** If the equation for Line 1 is $y = 8x$, which one could be the equation for Line 2?

  (i) $y = 6x$    (ii) $y = (1/2)x$    (iii) $y = 10x$    (iv) $y = 12x$

  **b.** Sketch the line $y = 4x$ in the same picture.

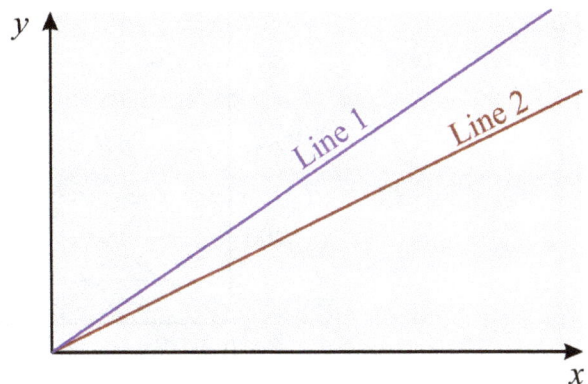

7. Andrew and his wife are buying a car, and they are deciding between two cars. Car 1 consumes 6.3 L of fuel per 100 km driven. The graph below is a distance-gasoline graph for Car 2.

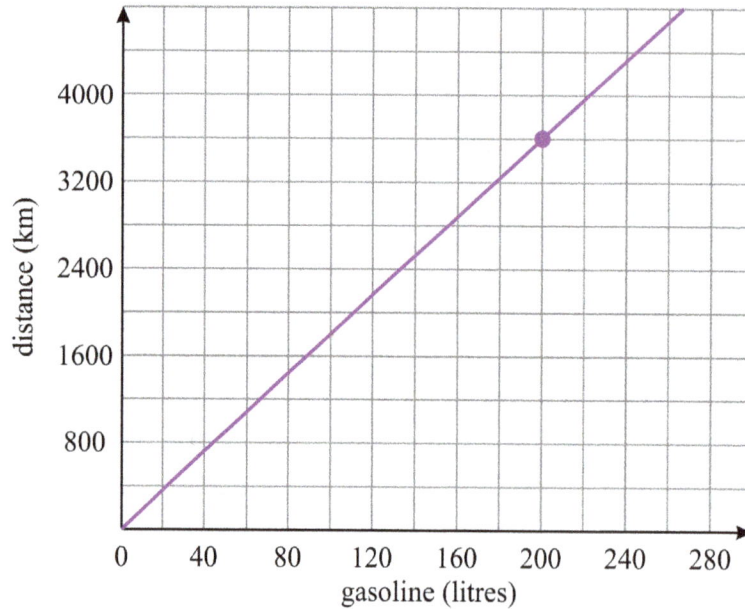

**a.** Which car has a better fuel efficiency?

**b.** Calculate how much more gasoline the car that guzzles more gas uses than the other car, over a distance of 1000 km (to the litre).

**c.** Estimating that they would drive 22 000 km in a year, and that gasoline costs $1.88 per litre, calculate how much they would save by purchasing the car that consumes less fuel. You can also make this calculation based on the current price of gas in your location.

**Puzzle Corner**

A lion can run at the speed of 72 km/h when chasing prey. The graph shows the distance that a cheetah covers, running after prey, over time. Which animal has a greater speed?

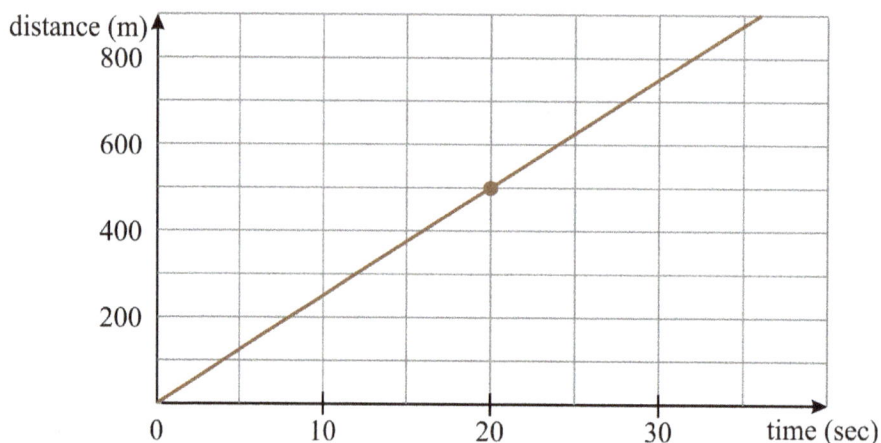

# Slope, Part 1

1. The illustrations show roofs with different amounts of pitch, or steepness.

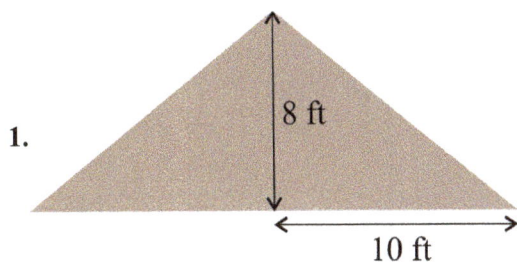

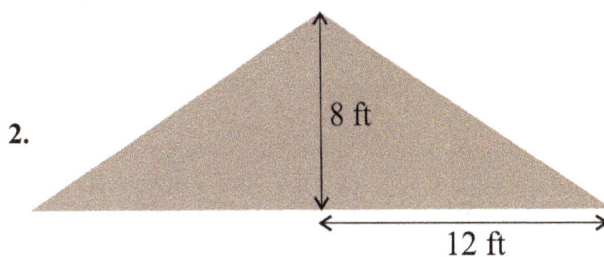

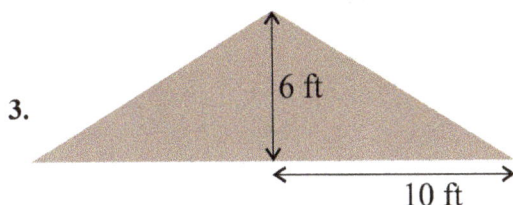

**1.**

8 ft

10 ft

**2.**

8 ft

12 ft

**3.**

6 ft

10 ft

**a.** Which roof is the steepest?

**b.** How could we use mathematics, and not visual inspection, to figure out which roof is the steepest?

2. Of these lines drawn in the grid, which one is the steepest? Which one is the least steep?

How could we determine that for sure? Try to think of a way, without peeking ahead.

Order the lines by steepness, the best you can.

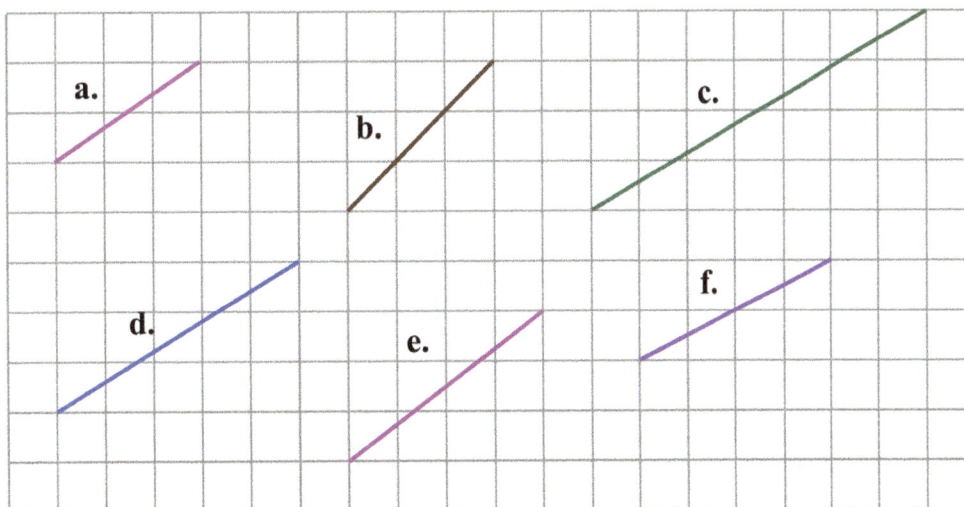

**a.**

**b.**

**c.**

**d.**

**e.**

**f.**

The **slope** of a line is a number that describes the steepness and direction of the line — its slant or inclination.

It is defined as the ratio $\dfrac{\text{change in } y\text{-values}}{\text{change in } x\text{-values}}$, between any two points on the line.

We often call this ratio as "rise over run" (as a fraction, $\dfrac{\textbf{rise}}{\textbf{run}}$). You will soon see why.

**Example 1.** What is the slope of this line?

Choose any point on the line. Here, we chose (3, 3). From it, draw a horizontal segment (the "run") toward the positive $x$ direction (right). Here, the "run" is four units.

From the end of that segment, draw another line segment (the "rise") UP, until you meet the line again. Here, the "rise" is three units.

Then write the ratio $\dfrac{\text{rise}}{\text{run}} = \dfrac{3}{4}$. The slope is 3/4. For each four

units of horizontal distance, the line rises by three units.

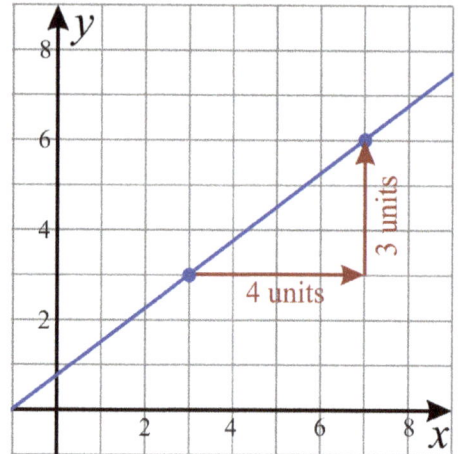

**Example 2.** If possible, it helps to choose two points on the line that are grid points. Then, the rise and the run will be whole numbers.

Here, slope = $\dfrac{\text{rise}}{\text{run}} = \dfrac{2}{6} = \dfrac{1}{3}$.

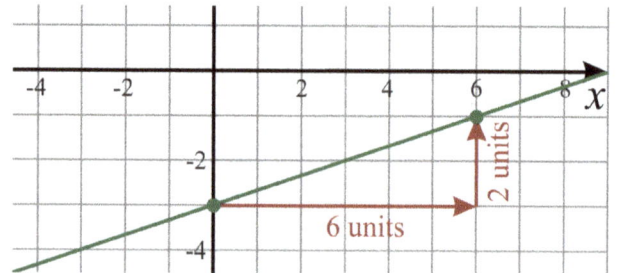

3. Find the slope of each line using the graph.

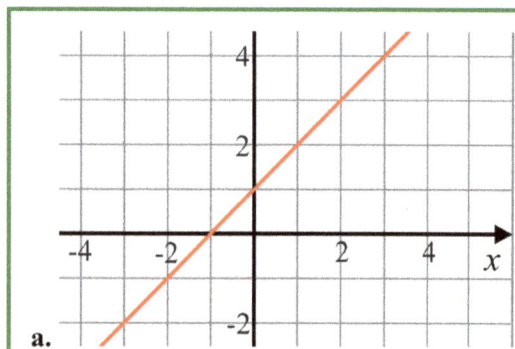

a.

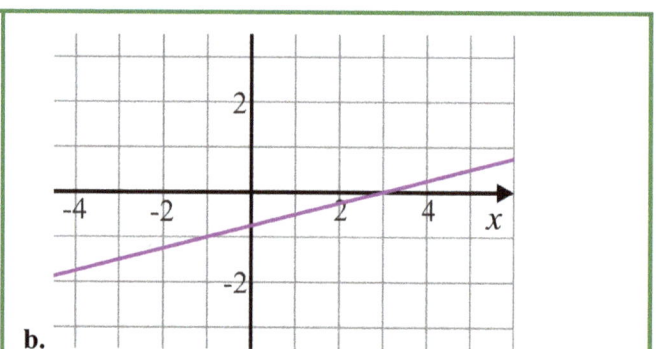

b.

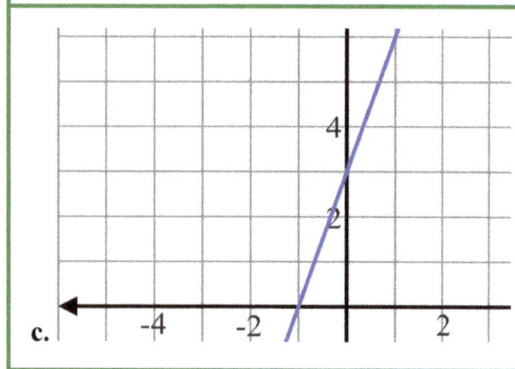

c.

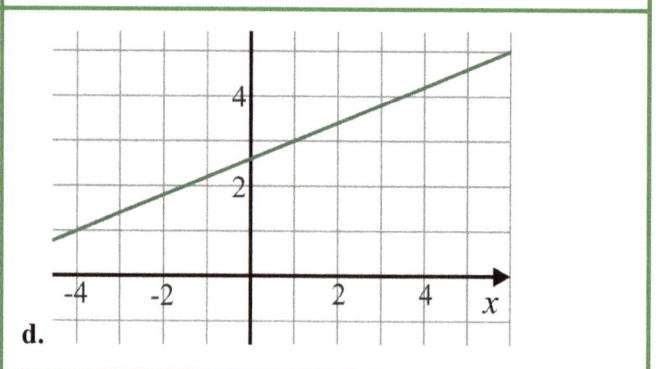

d.

4. Will it matter which points you use from the line? Will you get the same slope? Calculate the slope of this line

   **a.** using points A and B

   **b.** using points C and D

   **c.** using points A and D

We will look at a proof for this in the next lesson.

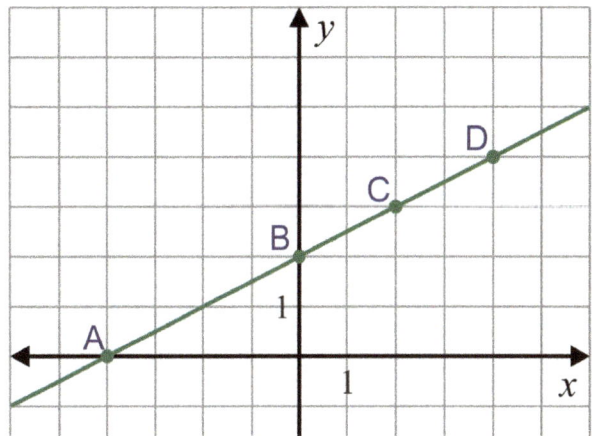

The slope can also be a negative number. In that case, we say that the line is **decreasing**: moving from left to right, it goes downwards.

Conversely, if the slope is positive, the line is **increasing**.

**Example 3.** What is the slope of this line?

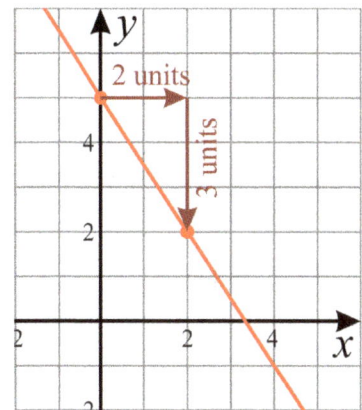

Again, we pick any point on the line and draw a horizontal line segment toward positive $x$. This time, to meet the line, we have to draw the second segment <u>downwards</u>. The vertical distance is 3 units, but it is "down", toward smaller $y$-values, which means **the rise is negative.**

The run is 2 units, and the rise is $-3$. So, the slope is $-3/2$.

5. Find the slope of each line using the graph.

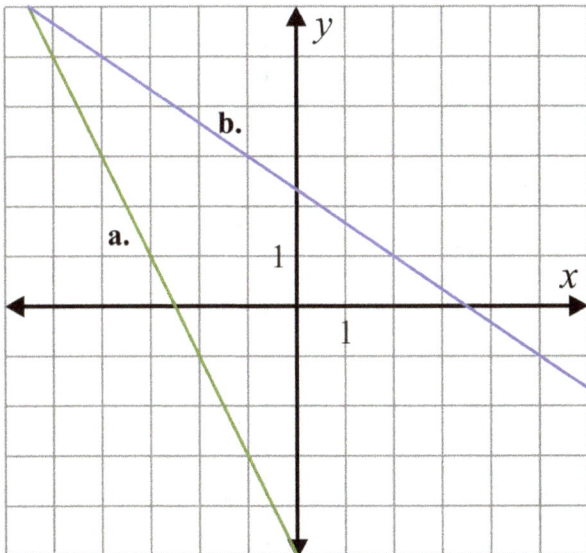

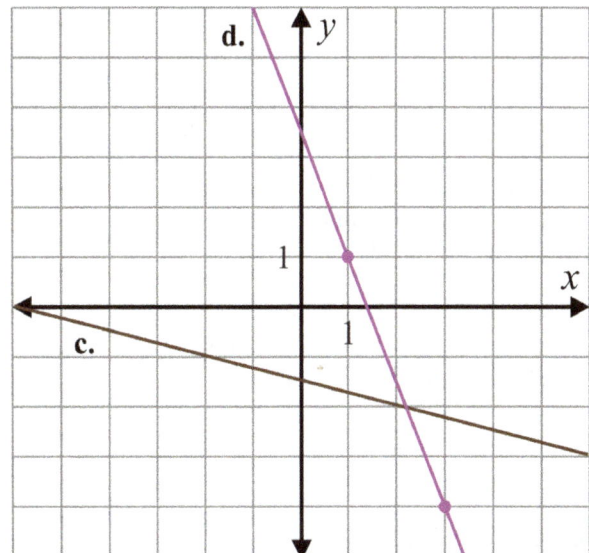

   **a.**

   **b.**

   **c.**

   **d.**

6. Find the slope of each line.

   **a.**

   **b.**

   **c.**

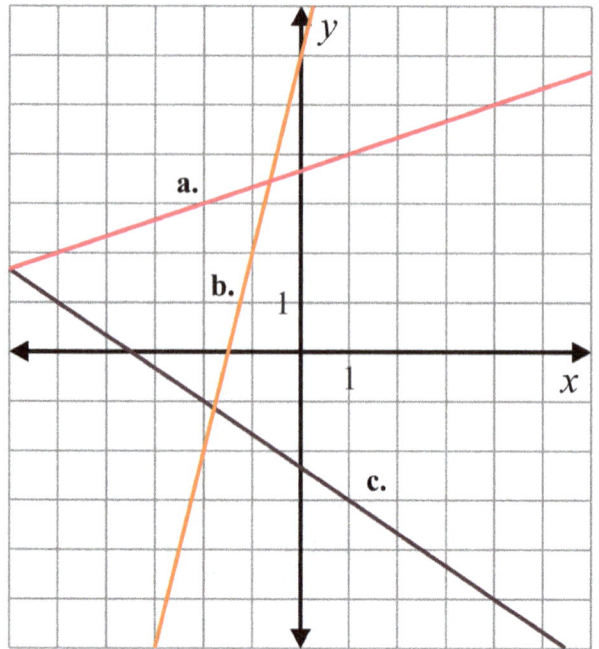

7. Find the slope of the line that goes through the given points:

   **a.** (0, 5)  and (5, 3)

   **b.** (−5, −4)  and (2, 1)

   **c.** (−4, 4)  and (5, −1)

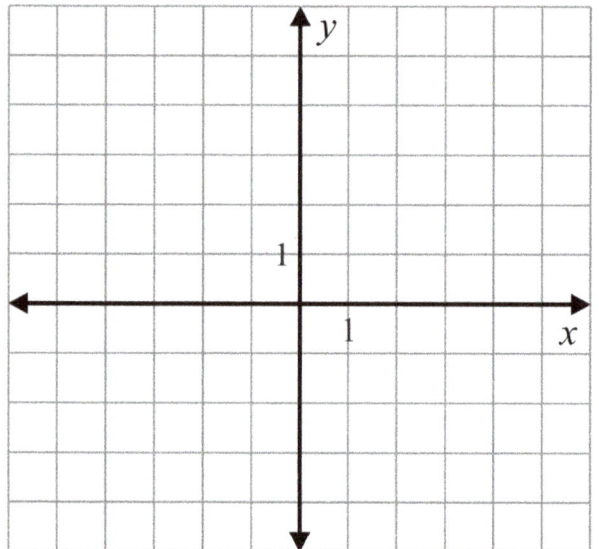

8. How can you find the slope of a line through two given points, without drawing the points?

   For example, find the slope of the line that goes through these points. You can check your work by graphing.

   **a.** (0, 2)  and (4, 5)

   **b.** (−3, 3)  and (2, −4)

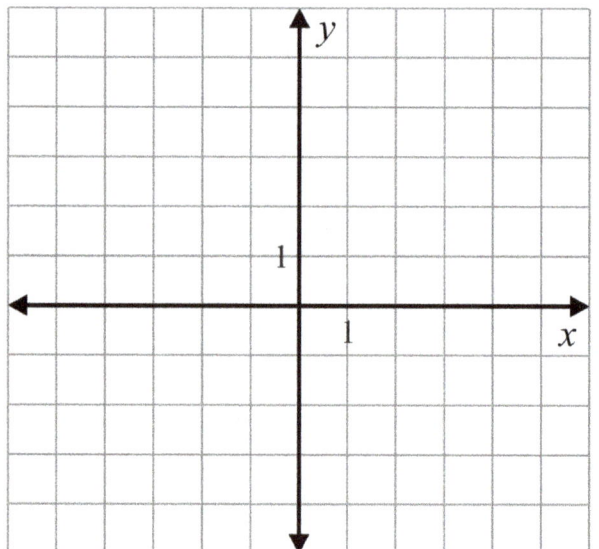

# Slope, Part 2

We will now prove that **it doesn't matter which two points on a line you use to calculate the slope**. Your task is to fill in some crucial points in the proof.

Let A, B, D, and E be different points on a line. See the illustration.

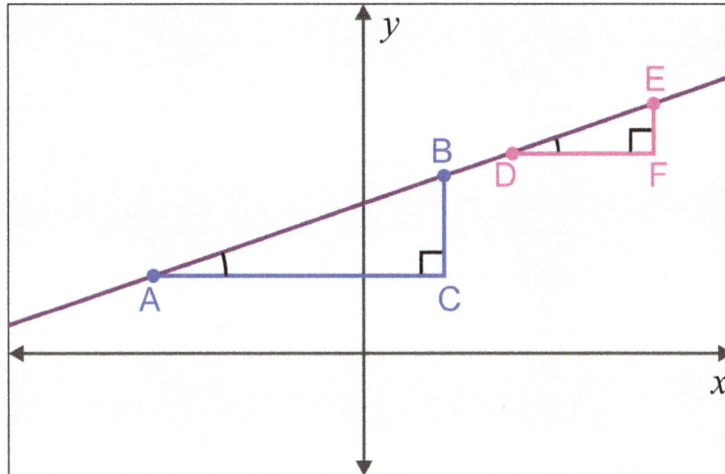

We will draw the line segments for the rise and run from point A to point C to point B. This creates the right triangle ABC.

Similarly, the rise and run from D to F to E creates the right triangle DEF.

What else is true about these two triangles?

Consider the angles BAC and EDF (marked with a single arc). The line segments AC and DF are

_____, and therefore, the angles BAC and EDF are _____ angles,

thus they are congruent.

Since ∠BAC = ∠EDF, and the angles BCA and EFD are equal (being right angles), this means the third

angles of the triangles ABC and DEF are equal too, and the triangles are _____.

In _____ triangles, corresponding side lengths are in the same ratio. More than that, the ratio of any one side length to another in *one* triangle equals the ratio of the corresponding sides in the other triangle.

Slope is the ratio rise/run. If calculated using points A and B, it is the ratio $\frac{BC}{AC}$.

If calculated using points D and E, it is the ratio $\frac{EF}{DF}$.

What can we therefore conclude?

We will now *derive* an equation for a line that goes through the origin, with slope $m$. This means we deduce it from known facts.

The illustration shows a generic line, with slope $m$, through the origin that goes through the first quadrant. Let A be any point on the line in the first quadrant, with coordinates $x$ and $y$. In other words, A is a generic point, symbolizing any point on the line (in the first quadrant).

We can draw a right triangle as you see in the illustration, where B is the point $x$ units from the origin on the horizontal axis. Then, the distance from B to A is $y$ units (because of the definition of coordinates).

Now, the slope of this line is $m$, and it is also rise/run,

or using the symbols in our illustration, it is $\dfrac{\phantom{xx}}{\phantom{xx}}$.

So, we can write the equation $m = \dfrac{\phantom{xx}}{\phantom{xx}}$.

Solving this equation for $y$, we get $y = $ _____ ,

Since we used a generic point on a line through origin, **this equation holds true for any point on the line in the first quadrant.**

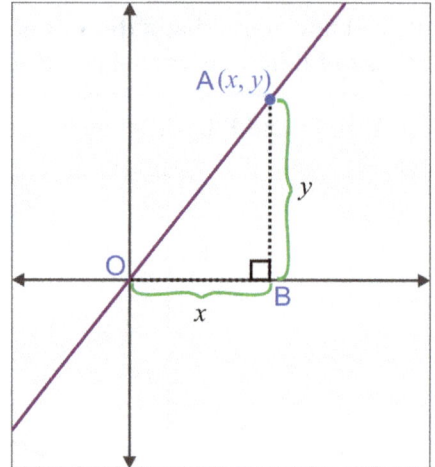

What about if the point is in the third quadrant, with negative coordinates?

In this situation, slope is positive, and both the run and the rise are positive numbers. Yet, the coordinates $x$ and $y$ are negative. We need to use the <u>absolute values</u> of $y$ and $x$ in calculating slope.

The slope is $m = \dfrac{\text{rise}}{\text{run}} = \dfrac{|y|}{|x|}$ .

However, the ratio $\dfrac{y}{x}$ would have the same value

— it would be positive, and equal $m$, since a negative divided by a negative is positive.

In other words, in this case, $\dfrac{|y|}{|x|} = \dfrac{\phantom{xx}}{\phantom{xx}} = $ _____ ,

And therefore, the equation $y = mx$ still holds.

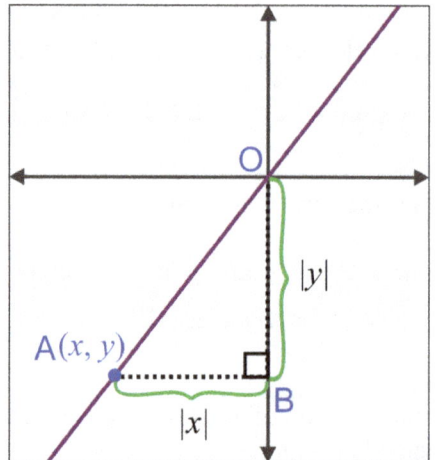

The case where the line is decreasing (goes through the 2nd and 4th quadrants) also involves dealing with absolute values, and is omitted here. The equation $y = mx$ still holds in those cases, too.

---

1. **a.** Find the equation of the line that goes through the origin and the point (20, 28).


   **b.** Find the equation of the line that goes through the origin and the point (−18, 14).

2. Find the coordinates of point B, using similar triangles.

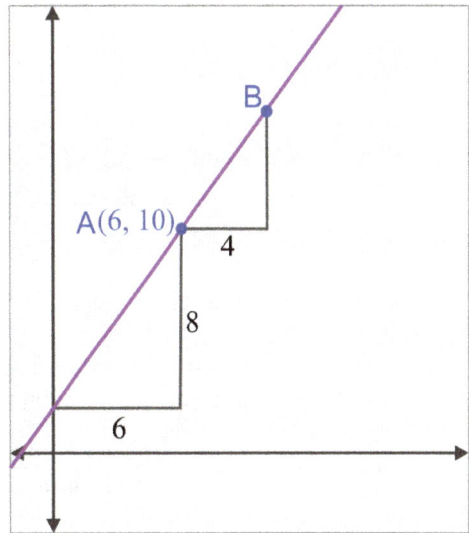

A(6, 10)
4
8
6
B

3. A line goes through the origin and the point (12, 10). The point (30, $b$) is on the same line. Find the value of $b$.

4. **a.** Draw a line that goes through the points A(5, 2) and C(15, 6).

   **b.** Calculate the slope of the line using points A and C. Also, draw the right triangle that shows the rise and the run.

   **c.** Plot point B on the line so that its $y$-coordinate is 3.

   **d.** Draw the rise-run segments between points A and B.

   **e.** Calculate the $x$-coordinate of point B.

   **f.** The two triangles formed are _____ triangles.

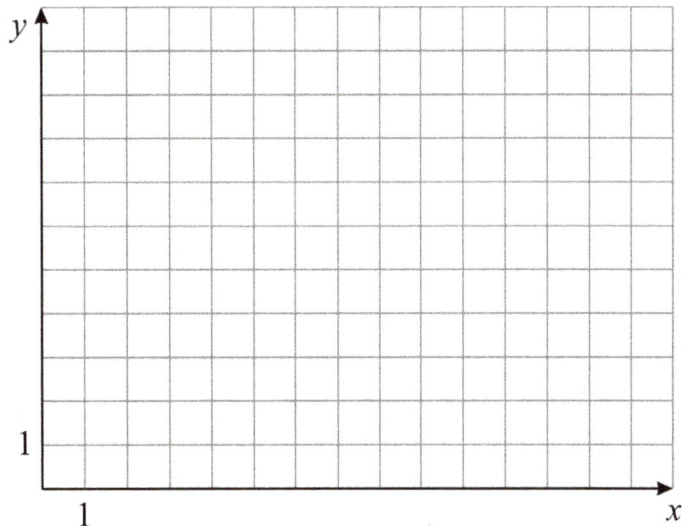

5. Do the three points (4, 3), (7, 5), and (10, 7) fall on the same line? How can you be sure?

6. **a.** A line goes through the points (−7, −4) and (7, 4). Write an equation for the line.

   **b.** Point ($s$, −5) is on that line. Find the value of $s$.

# Slope, Part 3

Slope can also be found from a table of values of a linear function. In that case, it is the same concept as the rate of change that you are already familiar with.

**Example 1.** The table gives values for a linear function.

| $x$ | −4 | −3 | −2 | −1 | 0 | 1 | 2 |
|---|---|---|---|---|---|---|---|
| $y$ | 13 | 11 | 8 | 7 | 5 | 3 | 1 |

Looking for patterns in the *x*-values and in the *y*-values, it is easy to see that as the *x*-coordinates increase by one, the *y*-coordinate*s* *decrease* by 2.

This means the slope is $\dfrac{\text{change in } y\text{-values}}{\text{change in } x\text{-values}} = \dfrac{-2}{1} = \mathbf{-2}$.

We will get the same by calculating the slope from any two points given in the table. For example, using the points (−3, 11) and (1, 3), we can see that the *y*-value decreases by 8, as the *x*-value increases by 4.

So, the slope is $\dfrac{\text{change in } y\text{-values}}{\text{change in } x\text{-values}} = \dfrac{-8}{4} = -2$.

1. Determine the slope of each line from the table of values. You can check your work by graphing.

a.

| $x$ | −3 | −2 | −1 | 0 | 1 | 2 | 3 |
|---|---|---|---|---|---|---|---|
| $y$ | −3.5 | −2 | −0.5 | 1 | 2.5 | 4 | 5.5 |

b.

| $x$ | −3 | −2 | −1 | 0 | 1 | 2 | 3 |
|---|---|---|---|---|---|---|---|
| $y$ | 2 | 0 | −2 | −4 | −6 | −8 | −10 |

c.

| $x$ | −4 | −2 | 0 | 2 | 4 | 6 | 8 |
|---|---|---|---|---|---|---|---|
| $y$ | 5 | 4 | 3 | 2 | 1 | 0 | −1 |

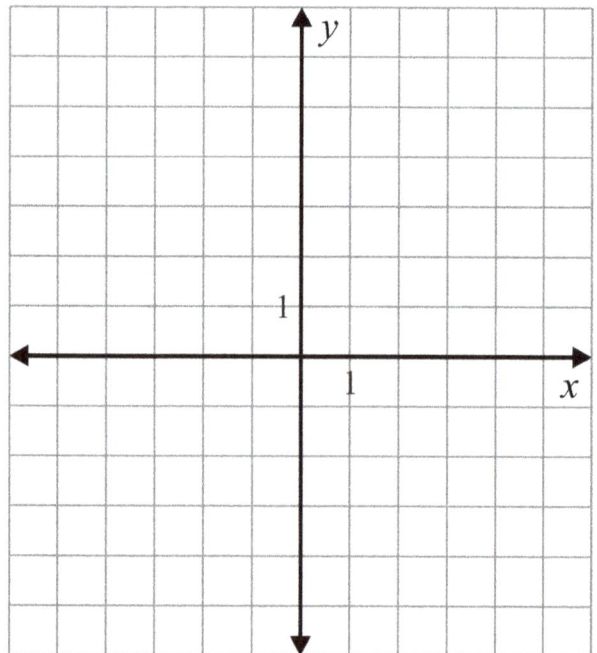

2. Find the slope of the line that goes through the points (−2, −3) and (0, 2).

3. Enrique calculated the slope of a line that goes through points (−4, 6) and (−2, 1) as follows:

$$\text{slope} = \frac{1 - 6}{(-4) - (-2)} = \frac{-5}{-2} = 2\,\tfrac{1}{2}$$

Find the error in his calculation.

4. **a.** Draw any two points on the horizontal line in the image.

   Now, calculate the slope using the coordinates of those points (change in *y*-values/change in *x*-values). What do you get?

   **b.** Draw any two points on the vertical line in the image.

   Now, *try* to calculate the slope using the coordinates of those points. What happens?

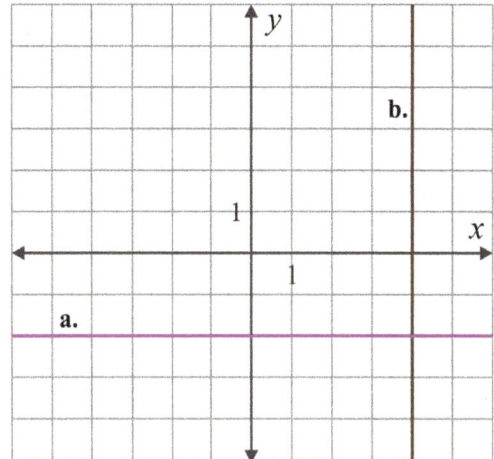

5. Find the slope of the line that goes through the given points. Also, graph the lines.

   **a.** (−3, 5) and (4, 5)

   Slope:

   **b.** (−2, 6) and (3, −4)

   Slope:

   **c.** (−5, 2) and (−5, −1)

   Slope:

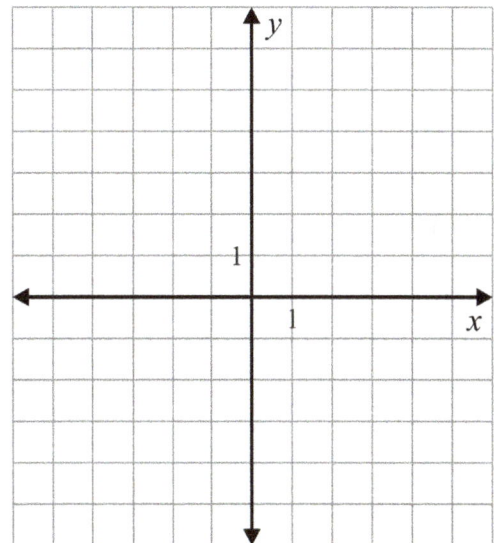

6. Determine the slope of each line. Notice carefully the scaling of the grids — it is not the same for each axis, but the way to find the slope is the same: rise over run.

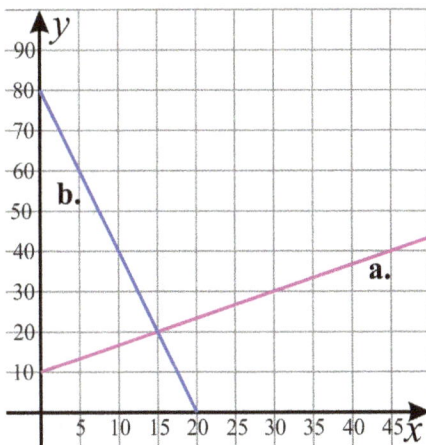

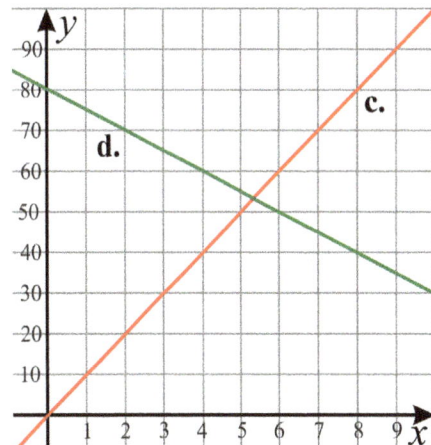

31

**Example 2.** Draw a line through (−2, 1) with a slope of 3/4.

Start out by plotting the given point. The slope of 3/4 means the rise is 3, and the run is 4. So, draw a horizontal "run" of four units from the given point. Then turn up, and draw a "rise" of 3 units. That is where you will have another point that is on the line — point (2, 4).

Now that you have two points, simply draw a line through them.

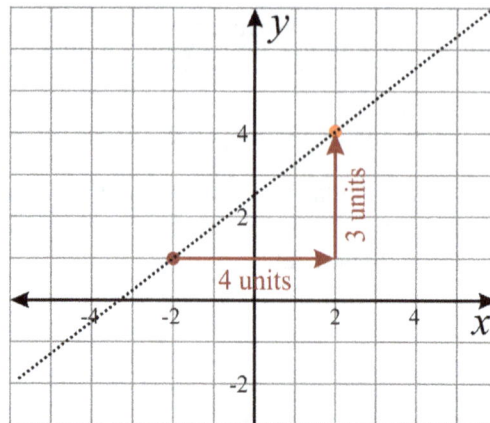

7. **a.** Draw a line with a slope of 2 that goes through the origin.

   **b.** Draw a line with a slope of 2/3 that goes through the point (−5, 1)

   **c.** Draw a line with a slope of 3/2 that goes through the point (−4, −4)

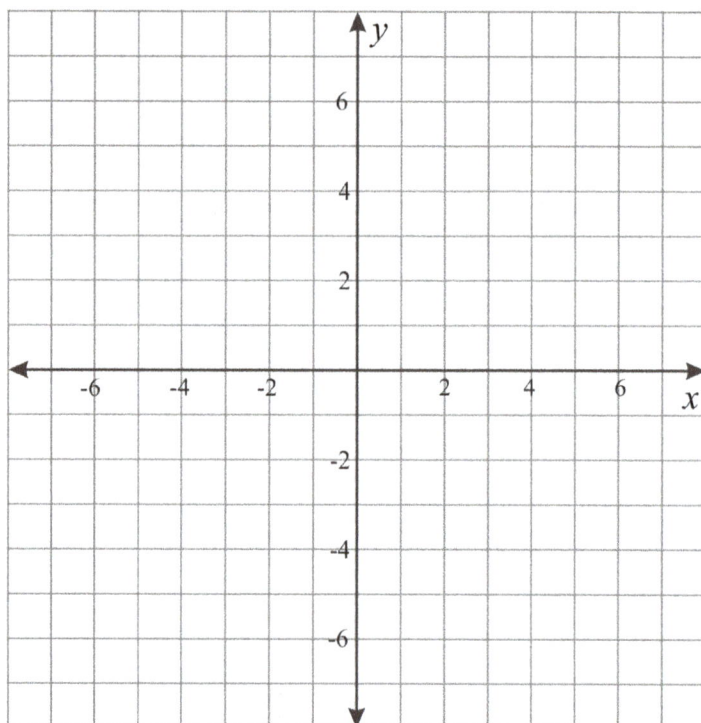

8. Draw two lines with a slope of 4/3. They can be drawn anywhere on the grid; they do not have to go through any specific point.

   Check: Your lines should be parallel.

9. Draw two lines with a slope of −4/3.

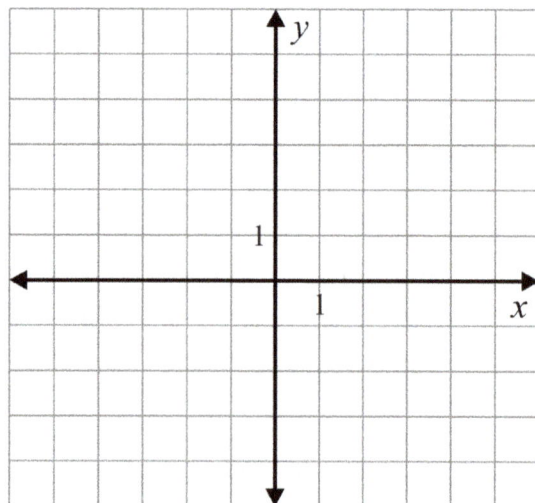

10. **a.** Draw a line with a slope of $-1/2$ and that goes through the point $(0, 6)$.

   **b.** Draw a line with a slope of $-3$ and that goes through the point $(-5, 6)$.

   **c.** Draw a line with a slope of $-2/3$ and that goes through the point $(0, 1)$.

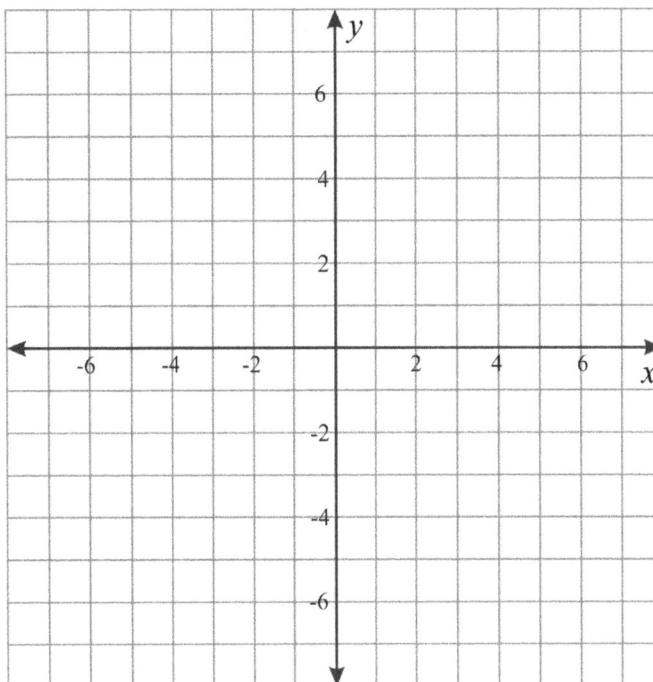

11. Draw any line with a slope of $30$.

12. Draw a line that goes through the point $(1, 70)$ and has a slope of $-15$.

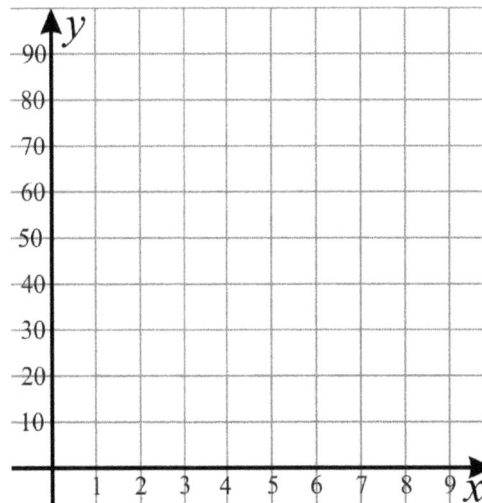

13. Alice tried to determine the slope of the line in this graph. She said, "The slope is 2, because the line goes through the point $(10, 20)$, and $20/10 = 2$."

   Explain what is wrong with her reasoning, how she can find the correct answer, and what that answer is.

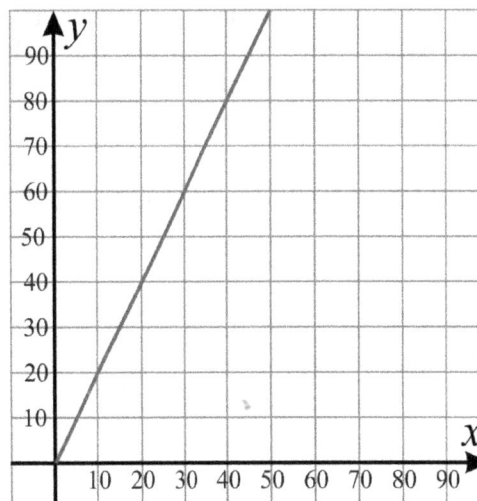

14. **a.** Draw a line with a slope of 1/5 that goes through the point (5, 4).

   **b.** Draw a line with a slope of −4 that goes through (−10, 6).

   **c.** Draw any line with a slope that is between 1 and 2.

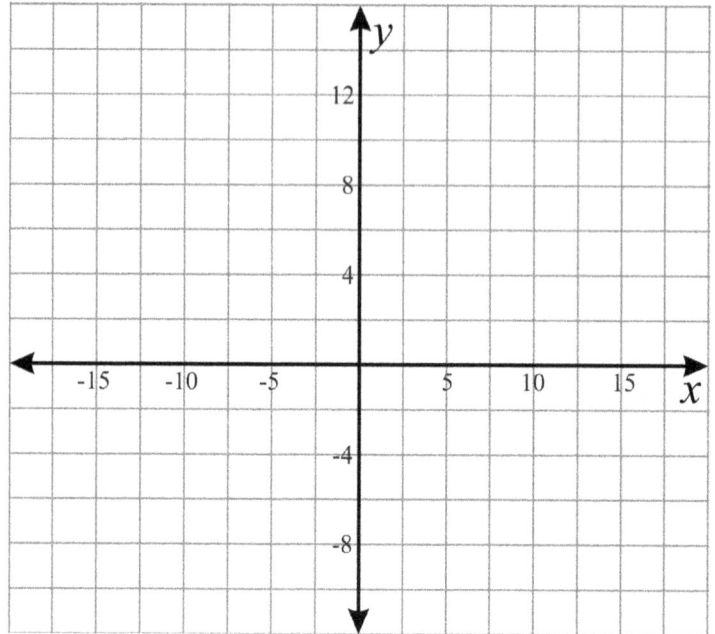

15. Three people travel the same distance of 10 km using three different methods: a car, a bicycle, or a moped. Their speeds are 75 km/h, 15 km/h, and 40 km/h, respectively.

   **a.** Graph the distance each vehicle covers as a function of time.

   **b.** How much faster is the car than the moped in covering the 10 km?

   **c.** How can you determine that from the graph?

   **d.** Draw in the grid a fourth line, to represent a fourth mode of transport that someone could use to travel this 10 km, and that is slower than the other three.

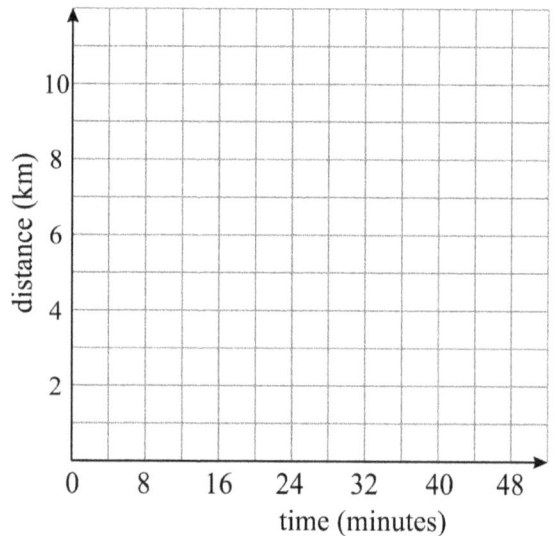

16. Do the three points (1, 3), (2, 7), and (4, 18) fall on one line? Explain.

17. Find $s$ so that the line through points (−15, 10) and ($s$, −5) has a slope of −3.

# Slope-Intercept Equation 1

1. Below, you see tables of values for two lines.
   The equation of Line 1 is given, and is $y = -2x$.

**Line 1:**

$y = -2x$

**Line 2:**

$y = $ _____

| x | y |
|---|---|
| −2 | 4 |
| −1 | 2 |
| 0 | 0 |
| 1 | −2 |
| 2 | −4 |
| 3 | −6 |

| x | y |
|---|---|
| −2 | 7 |
| −1 | 5 |
| 0 | 3 |
| 1 | 1 |
| 2 | −1 |
| 3 | −3 |

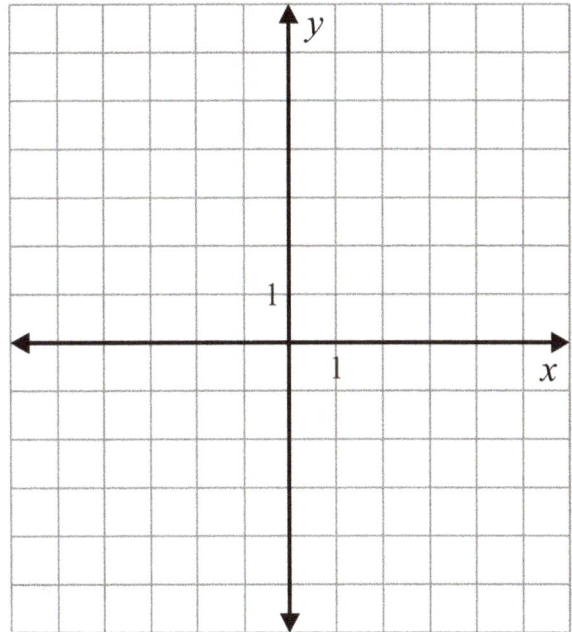

a. What is the slope of Line 1?     Of Line 2?

b. Graph both lines.

c. Where does Line 2 cross the $y$-axis?

d. Now compare the $y$-values <u>in the tables</u>. How do those $y$-values differ from each other?

   How can we see that same difference in the two graphs? In other words, what geometric transformation can you use to transform the first line to the second?

e. Write an equation for Line 2, of the form "$y =$ something".

f. Line 3 has the equation $y = -2x - 2$. How do the $y$-values of that line differ from those of Line 1?

   Therefore, what geometric transformation can you use to transform Line 1 to Line 3?

2. Line L is a line through the origin, with slope $m$, and its equation is therefore $y = mx$.

   a. What geometric transformation can you use to transform line L so that it crosses the $y$-axis at the point $(0, b)$?

   b. How would you change the equation $y = mx$ to reflect that change?

**Every linear function** — a line in the coordinate grid — **has an equation of the form** $y = mx + b$ .

In this equation, $m$ is the slope. In the past, you have learned to call the parameter $m$ the rate of change. This is the same concept as the slope.

What is the difference? Slope is used in the context of <u>graphing</u> and only applies to <u>lines</u>. Rate of change is used more generally, and can also be applied to functions that are not linear.

You have learned that the parameter $b$ is the **initial value** — in other words, the value of the function when the independent variable is zero. In the context of graphing, we call it the **y-intercept**, because the line crosses the y-axis at the point $(0, b)$.

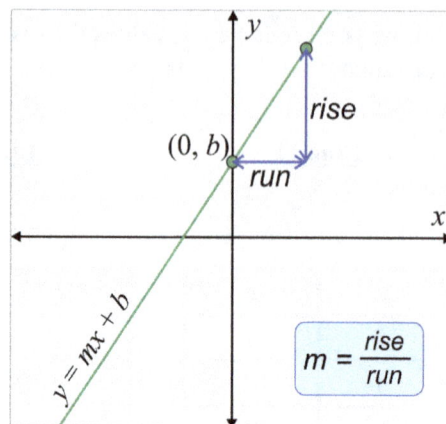

$$m = \frac{rise}{run}$$

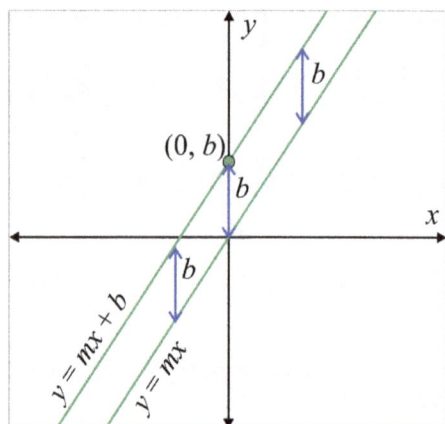

When comparing the lines $y = mx$ and $y = mx + b$, the latter line is located $b$ units above the former, if $b$ is positive, and $b$ units below it if $b$ is negative.

In other words, if we translate the line $y = mx$ by $|b|$ units up or down (up when $b > 0$ and down when $b < 0$), we get the line with the equation $y = mx + b$.

It is easy to plot a line when its equation is given in the slope-intercept form of $y = mx + b$. We can simply plot y-intercept point $(0, b)$, and then use the slope to find another point on the line.

**Example 1.** Plot the line $y = -\frac{1}{3}x - 1$.

The y-intercept is −1, so we draw the point $(0, -1)$.

The slope is −1/3. Draw the horizontal run as 3 units, then turn, and draw the "rise" as 1 unit going <u>down</u> (since the slope is negative).

Draw a point there.

Now, simply use those two points to draw a line through them.

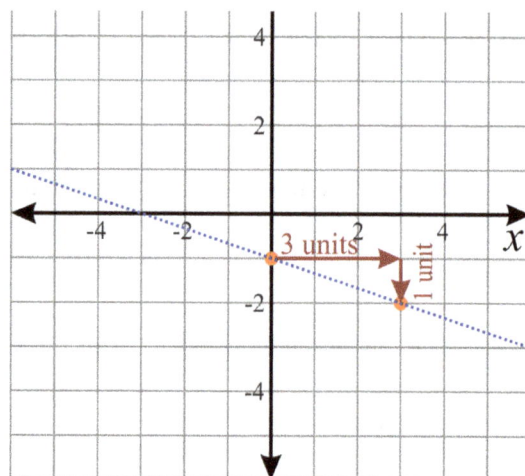

3. Graph the following lines.

  **a.** $y = x + 4$

  **b.** $y = -2x + 3$

  **c.** $y = 3x - 1$

  **d.** $y = -\dfrac{2}{3}x + 5$

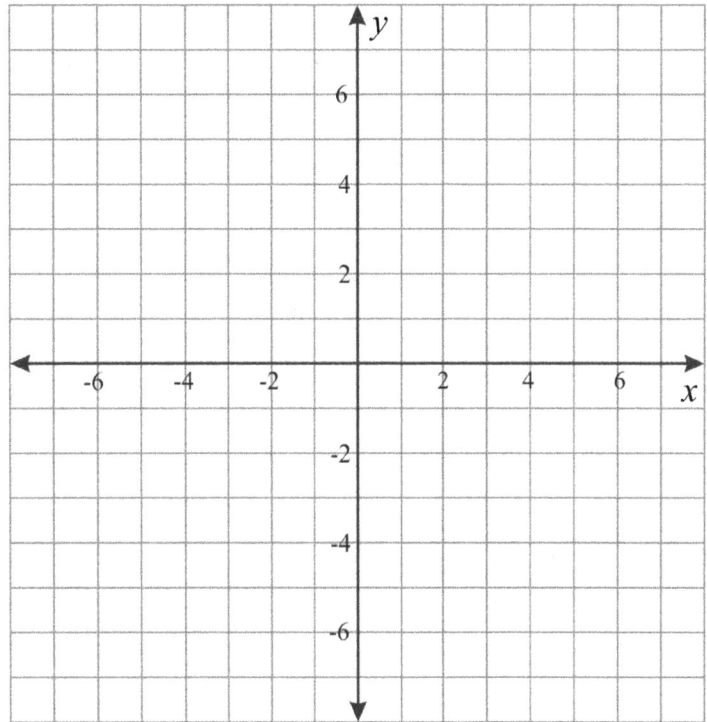

4. **a.** Graph the two linear functions
    $y = (2/3)x - 2$ and
    $y = (2/3)x + 1$.

  **b.** How do the $y$-values of these two linear functions differ?

  For example, when $x = 58$, what is the difference between the $y$-values of these two functions?

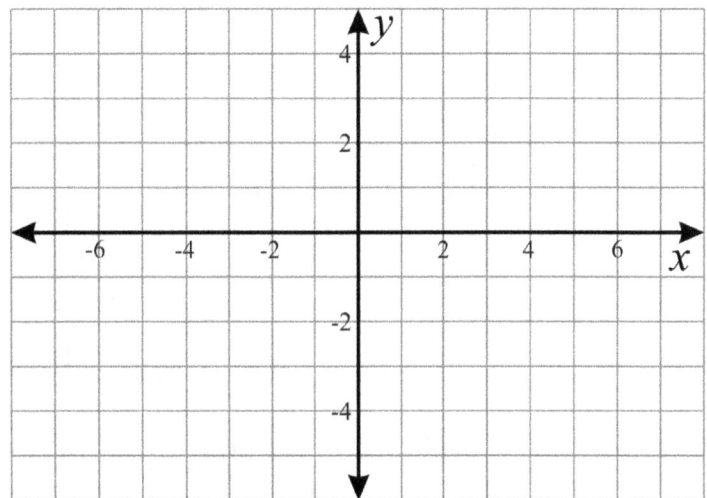

5. Graph the following lines.

  **a.** $y = \dfrac{2}{5}x + 1$

  **b.** $y = -\dfrac{1}{2}x - 2$

  **c.** $y = \dfrac{1}{3}x - 4$

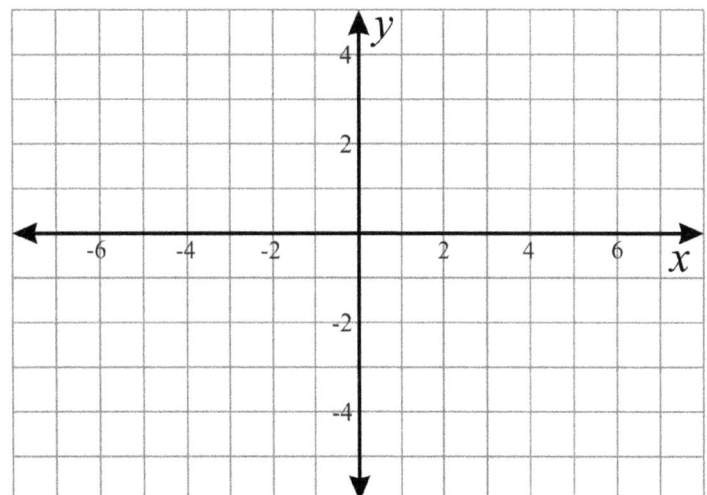

6. Graph the following lines.

   **a.** $y = -\frac{5}{2}x + 6$

   **b.** $y = \frac{5}{4}x - 5$

   **c.** a line with slope $-1$ and that goes through the point $(-2, -4)$

   **d.** a line with slope 4/3 and that goes through the point $(4, 6)$

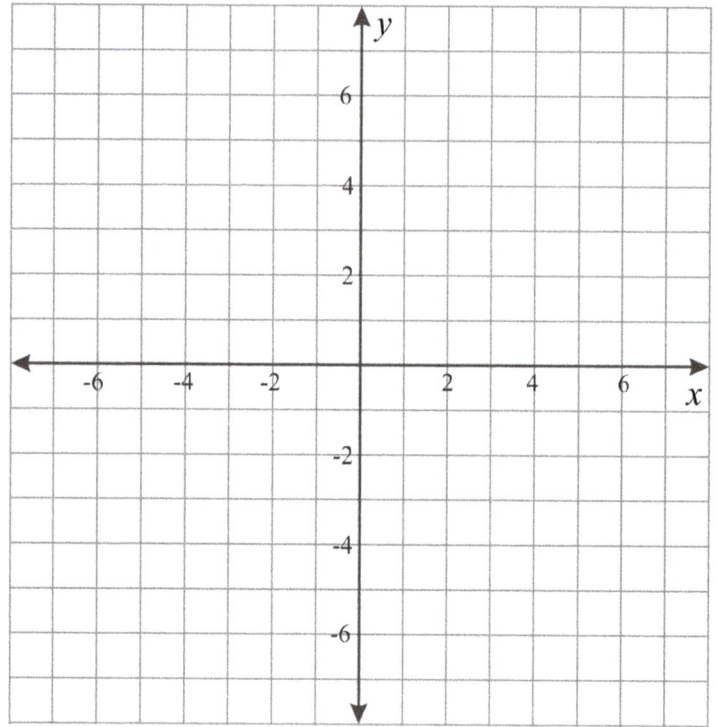

7. The tables of values give a list of points for four linear functions, or for four lines. Determine the equation of each, in slope-intercept form.

   **a.**

| $x$ | 0 | 10 | 20 | 30 | 40 | 50 | 60 |
|---|---|---|---|---|---|---|---|
| $y$ | 16 | 12 | 8 | 4 | 0 | −4 | −8 |

   **b.**

| $x$ | −15 | −10 | −5 | 0 | 5 | 10 | 20 |
|---|---|---|---|---|---|---|---|
| $y$ | −9 | −5 | −1 | 3 | 7 | 11 | 15 |

   **c.**

| $x$ | −24 | −20 | −16 | −12 | −8 | −4 | 0 |
|---|---|---|---|---|---|---|---|
| $y$ | −12 | −11 | −10 | −9 | −8 | −7 | −6 |

**a.** Determine the value of $a$ so that the line $y = ax + 3$ goes through the point $(2, -2)$.

**Puzzle Corner**

**b.** Determine the value of $b$ so that the line $y = 2x + b$ goes through the point $(-6, -3)$.

# Slope-Intercept Equation 2

**Example 1.** What is the equation of this line?

The line crosses the *y*-axis at (0, −2), so the **y-intercept is −2**.

The run is 3 and the rise is 1, so the **slope is 1/3**.

Now we simply put those two values into the generic equation of the line, $y = mx + b$.

The equation is $y = (1/3)x + (-2)$, which is usually simplified to .$y = (1/3)x - 2$.

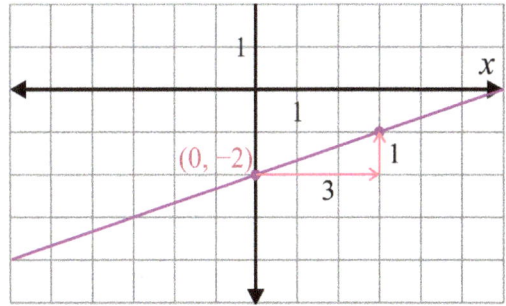

1. Write the equation for the line.

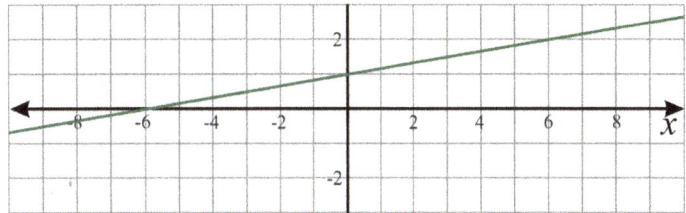

2. Write an equation for each line.

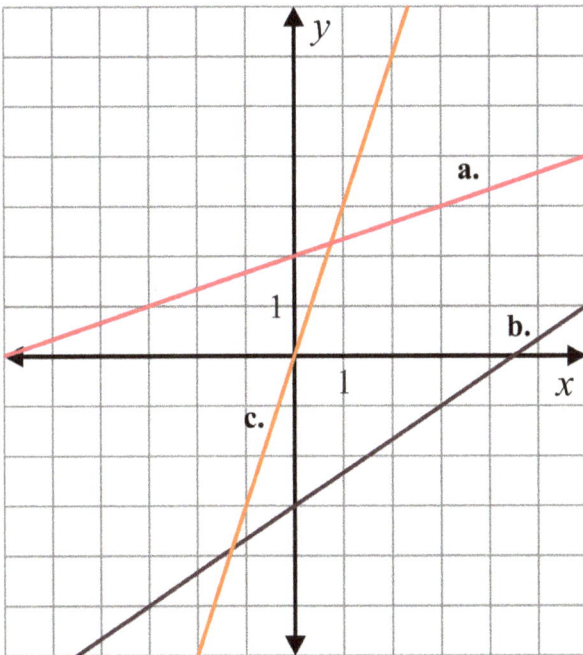

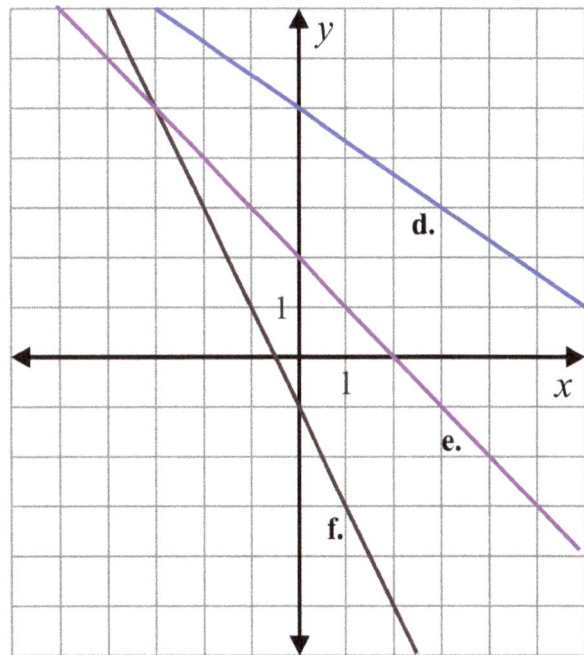

a.

b.

c.

d.

e.

f.

**Example 2.** A line goes through the points $(-4, 5)$ and $(7, -2)$. What is its slope?

The slope is $\dfrac{\text{change in } y\text{-values}}{\text{change in } x\text{-values}} = \dfrac{-2 - 5}{7 - (-4)} = \dfrac{-7}{11} = -\dfrac{7}{11}$.

The tricky part is to remember to always take the $x$ and $y$ coordinates in the same order (always write the *2nd* one minus the *1st* one). Checking by graphing can help.

3. Calculate the slope using two given points. Give your answer as a fraction. Sketching a graph can help.

   **a.** $(-4, 5)$ and $(6, -3)$

   **b.** $(20, 42)$ and $(85, 13)$

   **c.** $(-13, -7)$ and $(-6, 2)$

   **d.** $(-20, 13)$ and $(-5, -5)$

4. Determine the equation for each line, and graph the line.

   **a.**

   | $x$ | −2 | 0 |
   |-----|-----|---|
   | $y$ | −2.5 | 1 |

   **b.**

   | $x$ | −3 | 0 |
   |-----|-----|---|
   | $y$ | 8 | 5 |

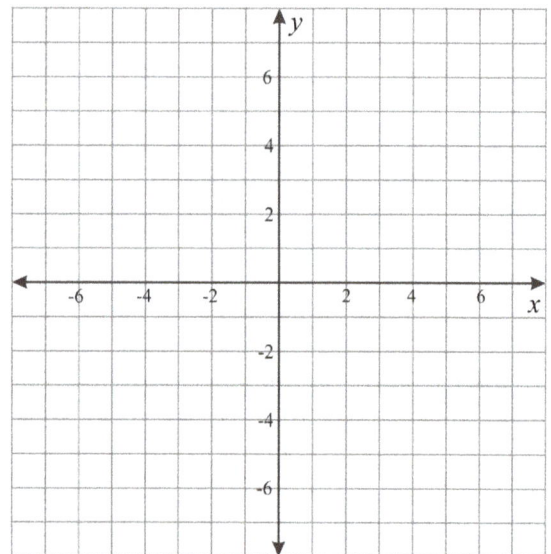

5. The value of a car depreciates (diminishes) over time. When new, Cindy's car cost $28 000. At 3 years, its value was $16 300.

   **a.** Assuming the value is changing in a linear fashion, write an equation to depict the car's value over time.

   **b.** What is the car's expected value at 5 years of age?

   **c.** What is the car's expected value at 10 years of age (surprising?). This shows us that in reality, the value does not follow a linear function.

> • Recall that a **linear function** can be written in the form $y = mx + b$.
>
> • A **proportional relationship** can be written in the form $y = mx$.

6. **a.** What is the main difference between the graph of a proportional relationship and a graph of a linear relationship?

   **b.** Is a proportional relationship always a *linear* relationship?

7. For both functions depicted below, tell whether the graph depicts a linear or a proportional relationship, or both. If proportional, tell the unit rate, and plot a point corresponding to it on the graph.

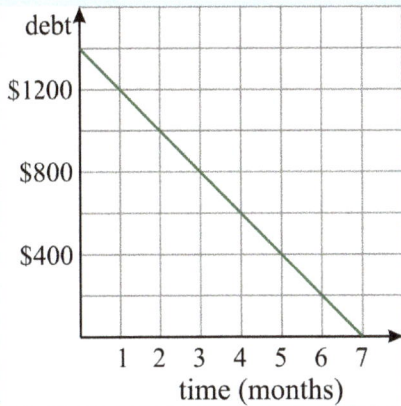

**a.**

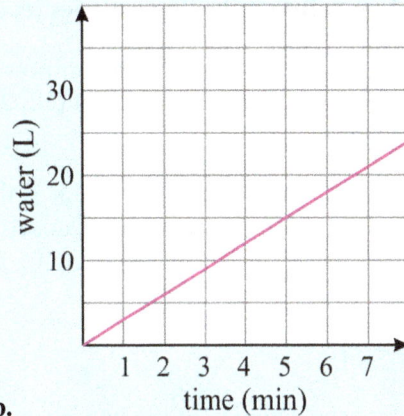

**b.**

8. Joe and Eric are taxi drivers. Joe charges $5, plus $2.25 per km driven. Eric charges $2.50 per km driven.

   **a.** Write an equation for the total cost as a function of number of kilometres driven, for each taxi driver.

   **b.** Is either function depicting a proportional relationship?

   **c.** Graph both functions (as lines).

   **d.** Where do the lines meet?

   What does that point signify?

   **e.** Fill in.

   For distances from 0 to _____ km,

   _____'s service is the better buy.

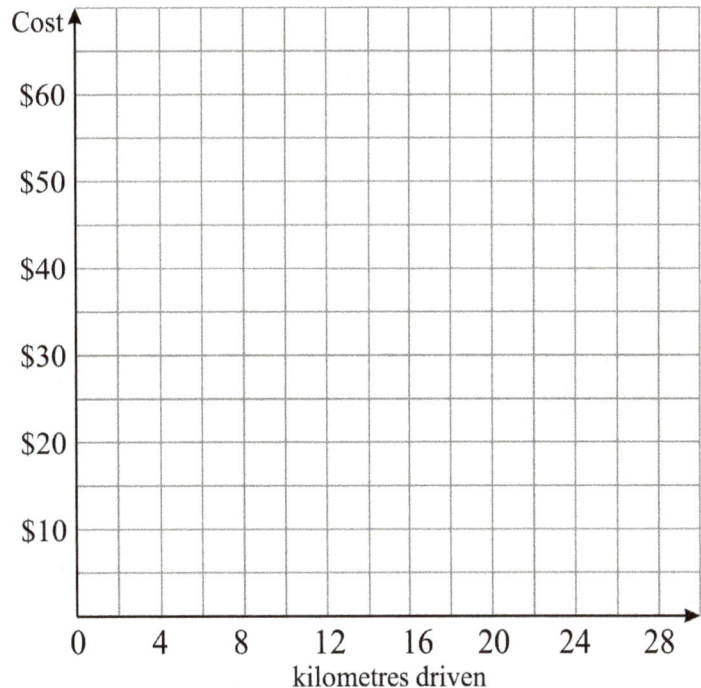

# Write the Slope-Intercept Equation

**Example 1.** What is the equation of a line that passes through (12, 25), with a slope of 3?

(Feel free to try to solve this problem on your own, before reading the solution below!)

The slope-intercept equation of a line is $y = mx + b$. Here we know the slope, but not the $y$-intercept. How can we find it out?

If you had a big sheet of paper, one way would be to continue the line, until it hits the $y$-axis. But a more efficient method is to use this fact:

  • **Any point on the line satisfies its equation.**

The point (12, 25) satisfies the equation $y = mx + b$. We know $m$ is 3, so actually we can say that the point (12, 25) satisfies the equation $y = 3x + b$.

Once we substitute 12 for $x$ and 25 for $y$, the equation will only have one unknown, $b$, and therefore we can solve it for $b$! So, let's do that:

$$25 = 3 \cdot 12 + b$$
$$25 = 36 + b$$
$$b = -11$$

This means the equation of the line is $y = 3x - 11$.

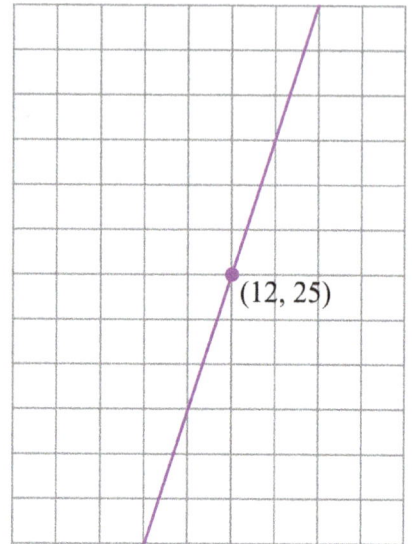

(12, 25)

1. Does the given point fall on the given line? Check by substituting its coordinates to the equation.

   **a.** (1, 5)  and  $y = -4x + 1$

   **b.** (6, −8)  and  $y = (-2/3)x - 4$

2. Find the equation of the line.

| **a.** passes through (20, 7) and has the slope 1/2 | **b.** passes through (−6, 5) and has the slope −2 |
|---|---|
| **c.** passes through (33, 40) and has the slope 1/3 | **d.** passes through (−20, −30) and has the slope −2/5 |

3. Find the equation of each line.

a.
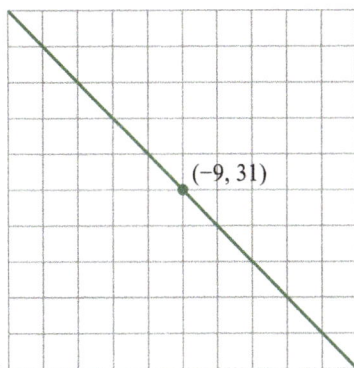
(−9, 31)

b.
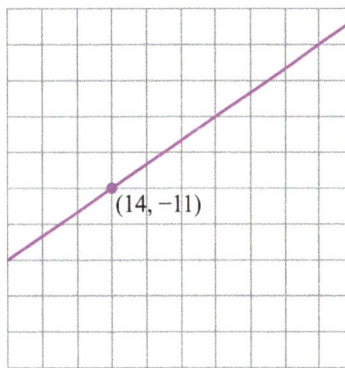
(14, −11)

c.
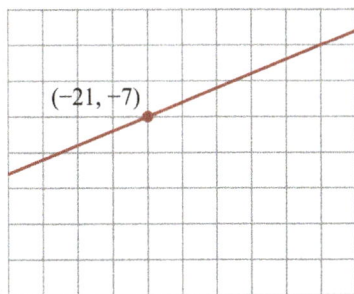
(−21, −7)

d.
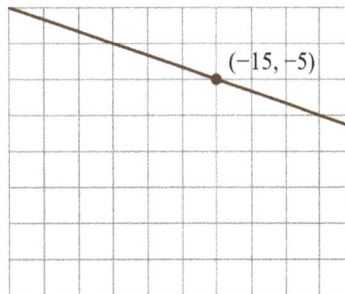
(−15, −5)

4. Match the descriptions and the equations.

**a.** Passes through (−2, 4) and has slope −2

**b.** Passes through (0, −2) and has slope −2

**c.** Passes through (1, −2) and (½, 0)

**d.** Passes through (−½, −4) and (1, 2)

**(i)** $y = -4x + 2$

**(ii)** $y = -2x - 2$

**(iii)** $y = 4x - 2$

**(iv)** $y = -2x$

5. An overseas mail service has a fixed fee, plus a fee based on the weight of the package. Kayla mailed two packages. One weighing 2 kg cost her $97, and another weighing 1.6 kg cost her $83.40.

 **a.** Write an equation for the cost (C) of sending a package that weighs $w$ kg.

 **b.** How much would sending a 1-kg package cost?

6. Eric sells home-made sauerkraut in his neighbourhood. He charges a fixed delivery fee, plus $6.25 per jar. Louise bought three jars, and it cost her $23.50.

 **a.** What would it cost to purchase seven jars of sauerkraut from Eric?

 **b.** Write an equation for the cost of $t$ jars of sauerkraut.

7. Lisa has been adding $140 to her savings account every month for a while (and not using the account otherwise). After 13 months of doing that, her account balance was $2710.

 **a.** What was her balance 13 months ago?

 **b.** Write an equation to model the total in her savings account, if she continues like this.

 **c.** How much will she have saved in 3 years?

---

A boat started travelling across the sea with a constant speed, but after several hours of travelling, it had to change to a slower (constant) speed. The table shows the distance the boat had covered after so many hours.

**Puzzle Corner**

What is this slower speed?

After how many hours of travelling did the boat change to a slower speed?

| Time (hours) | Distance (km) |
|---|---|
| 4 | 128 |
| 6 | 192 |
| 13 | 400 |
| 20 | 596 |

# Horizontal and Vertical Lines

You will recall that a horizontal line has a slope of zero, and a vertical line has no slope. What kind of equation can we write for these types of lines, if any?

A horizontal line follows the slope-intercept form of $y = mx + b$. We substitute $m = 0$, and get the generic equation $y = b$. Each point on the line has the $y$-coordinate $b$, no matter what the $x$-coordinate is.

Similarly, a vertical line has the equation of the form $x = a$.
Each point on the line has the $x$-coordinate of $a$.

**Example 1.** What are the equations for Line 1 and Line 2 in the image on the right?

For points on Line 1, the $x$-coordinate stays the same, and is always 5. So, the equation is $x = 5$.

For points on Line 2, the $y$-coordinate stays the same, and is always $-3$. So, the equation is $y = -3$.

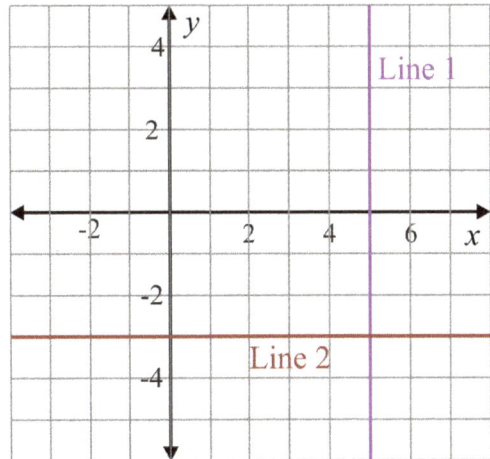

1. Draw the following lines.

   **a.** $x = -4$

   **b.** $y = 6$

   **c.** $x = 1$

   **d.** $y = -2$

   **e.** $y = x + 2$

   **f.** $y = -x + 2$

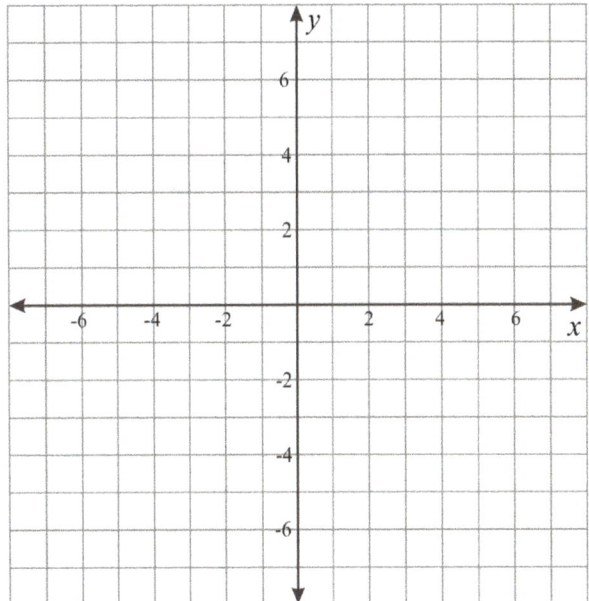

2. Write the equation of each line.

   **a.** passes through $(-14, 12)$ and is vertical

   **b.** passes through $(-60, 30)$ and horizontal

3. Write the equation of each line.

| **a.** passes through (−34, 65) and (−34, −80) | **b.** passes through (−15, 12) and has the slope −2/3 |
| --- | --- |
| **c.** passes through (100, −90) and (35, −90) | **d.** passes through (10, −90) and (−40, 50) |
| **e.** has slope zero, and passes through (3, 6) | **f.** has no slope, and passes through (3, 6) |

4. Find the area of the rectangle bounded by the lines $y = 300$, $y = -150$, $x = 20$, and $x = -70$. You can use the grid to help, or to help check your work, but the answer can be found without it!

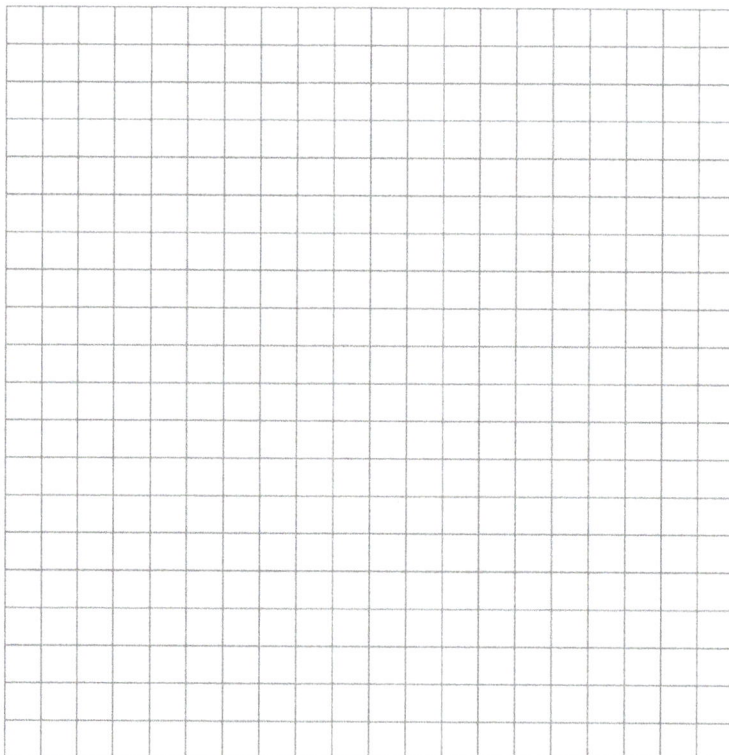

5. Match the descriptions and the equations.

**a.** Passes through (2, −4) and has slope 1/4

**(i)** $y = (1/4)x - 9/2$

**b.** Passes through (−½, ½) and has slope 0

**(ii)** $y = (1/4)x$

**c.** Passes through (−8, −2) and (0, 0)

**(iii)** $x = 5$

**d.** Passes through (5, −4) and (5, 2)

**(iv)** $y = ½$

6. Find the area of the triangle bounded by the lines $y = 6$, $x = 6$, and $y = (-3/2)x + 9$.

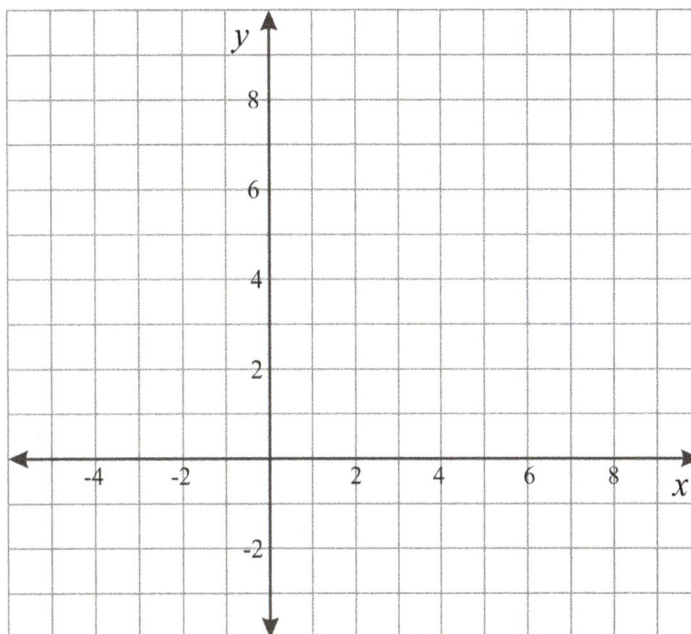

7. The image shows three lines. Find the equation of each. The gridlines are one unit apart.

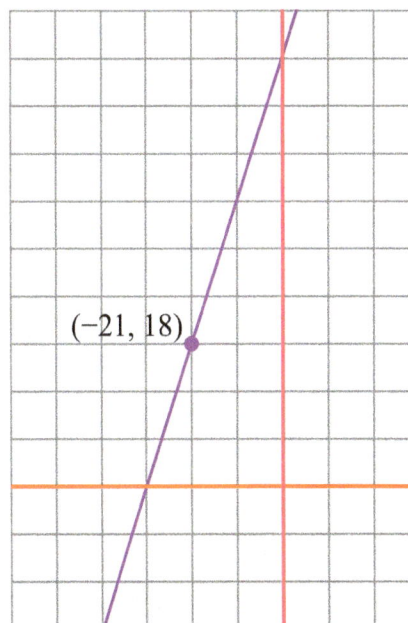

(−21, 18)

47

# The Standard Form

The **standard form** of an equation for a line is $\mathbf{A}x + \mathbf{B}y = \mathbf{C}$, where A, B, and C are integers, and A is nonnegative. For example, $11x - 9y = -4$ is in standard form.

Any of the numbers A, B, or C can be zero. For example, the equation $x = 4$ is in the standard form. (What special kind of line is it?)

**Example 1.** The equation $2x + 4y = 5$ is in standard form. Write it in slope-intercept form, and graph it.

Writing it in the slope-intercept form means we solve for $y$ in the equation $2x + 4y = 5$, since the slope-intercept form is in the form of "$y = $ (something)".

$$
\begin{aligned}
2x + 4y &= 5 \qquad &&\Big|{-2x} \\
4y &= 5 - 2x \qquad &&\Big|\div 4 \\
y &= (5/4) - (1/2)x
\end{aligned}
$$

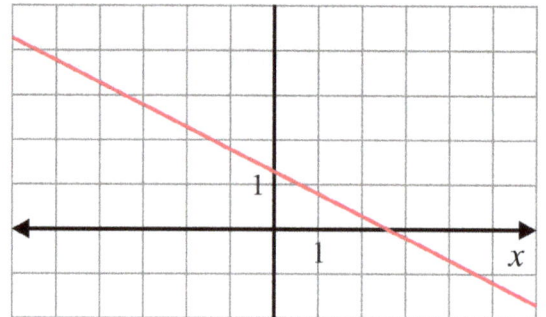

Lastly, we reorganize the right side of the equation so that the term with $x$ is written first:

$$y = -(1/2)x + (5/4)$$

Now it is easy to see that the slope is $-\frac{1}{2}$ and that the $y$-intercept is 5/4, and we can graph the line based on that.

1. Are the following equations in the standard form? For those that are not, transform them so that they are. All normal rules for manipulating equations apply! You can multiply both sides by something, for example.

   **a.** $3x + 6y = -8$

   **b.** $2y = -8 - 5x$

   **c.** $\dfrac{1}{4}x - \dfrac{3}{4}y = 5$

   **d.** $5x + y = 0$

   **e.** $y = 9$

   **f.** $-x + 2y = -6$

2. Transform each equation in standard form to the slope-intercept form.

| **a.** $2x - 4y = 5$ | **b.** $x + 2y = -10$ |
|---|---|
| **c.** $5x + 6y = -3$ | **d.** $3x + 3y = 7$ |

The standard form makes it handy to find $x$ and $y$-intercepts of a line. You already know that the $y$-intercept is the point where the line crosses the $y$-axis; in other words, it is the point on the line where $x = 0$.

Similarly, the $x$-intercept is the point where the line crosses the $x$-axis — the point where $y = 0$.

**Example 2.** Find the $x$ and $y$-intercepts of the line $3x + 4y = -6$.

To find the $x$-intercept, we set $y$ to zero in the equation:  $3x + 4 \cdot 0 = -6$ , from which $3x = -6$, and $x = -2$. The line crosses the $x$-axis at $x = -2$.

To find the $y$-intercept, we set $x$ to zero:  $3 \cdot 0 + 4y = -6$ , from which $4y = -6$, and $y = -3/2$. The line crosses the $y$-axis at $y = -3/2$.

Finding the $x$ and $y$-intercepts gives us two points: $(-2, 0)$ and $(0, -3/2)$, and it is easy to draw the line based on those:

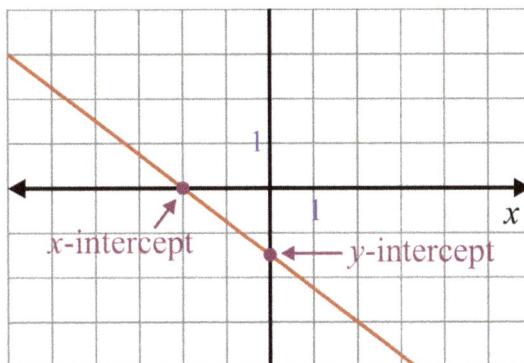

3. For each line given in standard form, find its $x$ and $y$-intercepts. Then graph the line.

   **a.** $3x - y = 6$

   **b.** $5x + 2y = -10$

   **c.** $4x - 3y = -12$

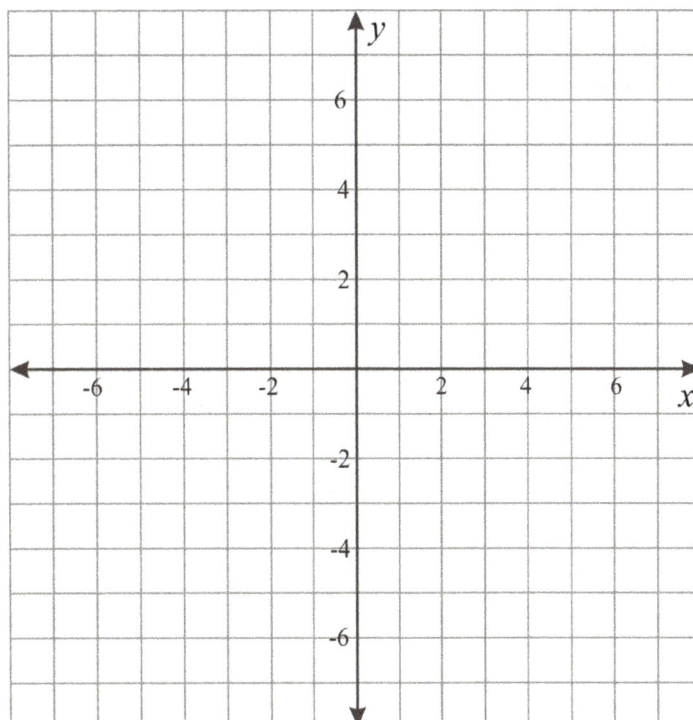

49

4. **a.** Is an equation for a horizontal line, such as $y = 11$, in the standard form?

In the slope-intercept form? If yes, give the slope.

**b.** Is an equation for a vertical line, such as $x = -70$, in the standard form?

In the slope-intercept form? If yes, give the slope.

5. Find the equation of the line, and give it in the standard form.

| **a.** has slope 3, and contains (2, 9) | **b.** passes through (1, −5) and (8, −4) |
| --- | --- |
| | |

6. Is the given point on the given line? If yes, graph the point and the line, and give the equation of the line in standard form.

**a.** (2, −4) and $-2x - y = 1$

**b.** (0, 6) and $\dfrac{3}{4}x - \dfrac{1}{2}y = -3$

**c.** (12, 3) and $\dfrac{1}{2}x - 2y = 0$

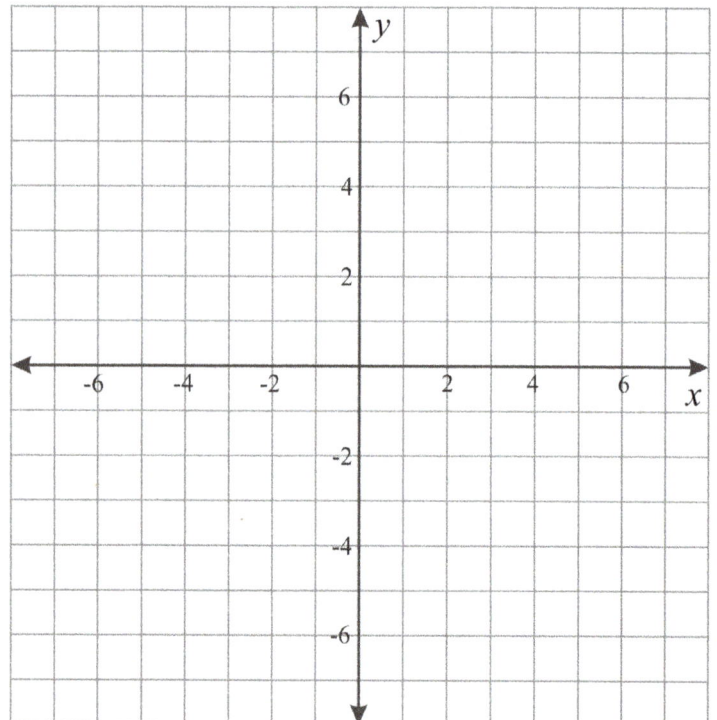

# More Practice
## (This lesson is optional.)

1. Transform each equation to the standard form.

| | |
|---|---|
| **a.** $y = 5x - 3$ | **b.** $y = x/2 + 3$ |
| **c.** $-\dfrac{1}{3}x - \dfrac{1}{6}y = 1$ | **d.** $\dfrac{2}{5}x = \dfrac{3}{4}y - 1$ |

2. For each line given in standard form, find its $x$ and $y$-intercepts. Then graph the line.

   **a.** $5x - 4y = 10$

   **b.** $10x + 4y = -20$

   **c.** $x - 3y = -6$

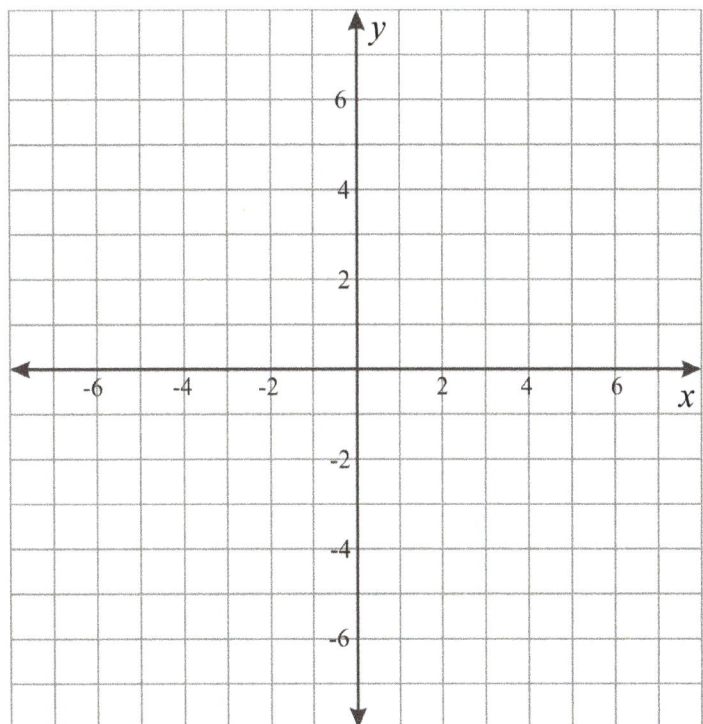

3. Find the equation of the line, and give it in the standard form.

| **a.** has slope 4, and contains $(5, 2)$ | **b.** passes through $(-1, 4)$ and $(1, -6)$ |
|---|---|
| **c.** is vertical and contains $(-3, 11)$ | **d.** is horizontal and contains $(9, -4)$ |
| **e.** has slope $-1/2$, and contains $(3, 11)$ | **f.** passes through $(-5, -3)$ and $(-2, 9)$ |

4. Transform each equation to the slope-intercept form.

| **a.** $y - 2 = 2(x - 1)$ | **b.** $9x = 7 - 4y$ |
|---|---|
| **c.** $3x + 9y = -2$ | **d.** $\frac{2}{3}x + \frac{1}{2}y = -1$ |

# Parallel and Perpendicular Lines

1. Graph the lines. Then determine which lines are parallel to each other, and which ones are perpendicular to each other.

   **a.** $y = -3x + 2$

   **b.** $y = (-1/3)x - 2$

   **c.** $y = (1/3)x + 1$

   **d.** $y = -3x - 2$

   **e.** $y = (-1/3)x + 5$

   **f.** $y = -3x - 1$

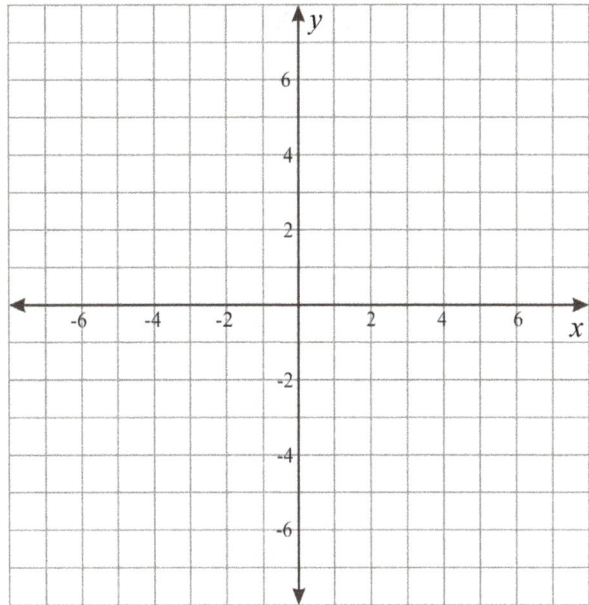

Based on your work above, and also what you have observed in the previous lessons, fill in:

**Lines that are parallel have the same _____.**

For two lines that are perpendicular, you may have noticed something special about their slopes, too.

If two lines are **perpendicular, the product of their slopes is −1.**

For example, slopes −4 and 1/4 signify lines that are perpendicular, because $-4 \cdot (1/4) = -1$.

To find the slope of the lines that are perpendicular to a line with known slope, take the opposite of the reciprocal of the known slope. For example, if the slope is 2/9, the slope of a perpendicular line is −9/2.

**Example 1.** A line passes through (7, −3) and is perpendicular to the line $y = 2x$. What is its equation?

The line $y = 2x$ has the slope 2. The opposite of the reciprocal of 2 is −1/2.

So, now we write the equation for a line with slope −1/2 and that passes through (7, −3). To do that, we substitute (7, −3) in the equation $y = -(1/2)x + b$, and solve for $b$:

$$-3 = -(1/2) \cdot 7 + b$$
$$-3 = -7/2 + b$$
$$b = 1/2$$

The equation of the line is $y = -(1/2)x + 1/2$.
Graphing both lines, we see everything looks good.

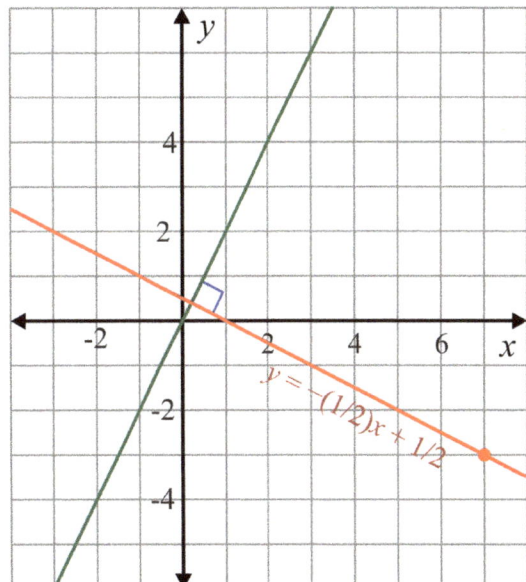

2. Find the opposite of the reciprocal of each number.

   **a.** 5                 **b.** −3/4            **c.** −6             **d.** 1/8

3. Which of the following lines are parallel? Which are perpendicular to each other?

   **a.** $y = \dfrac{3}{8}x - 3$      **b.** $y = -\dfrac{3}{8}x + 8$      **c.** $y = -\dfrac{8}{3}x + 8$      **d.** $y = -\dfrac{3}{8}x - 3$

4. **a.** Line L is plotted on the right. Line M is parallel to line L, and passes through the point $(-1, 0)$.

   Find the equation of Line M, in slope-intercept form, and plot it.

   **b.** Line N is perpendicular to line L, and passes through the point $(-1, 0)$.

   Find the equation of Line N, in slope-intercept form, and plot it.

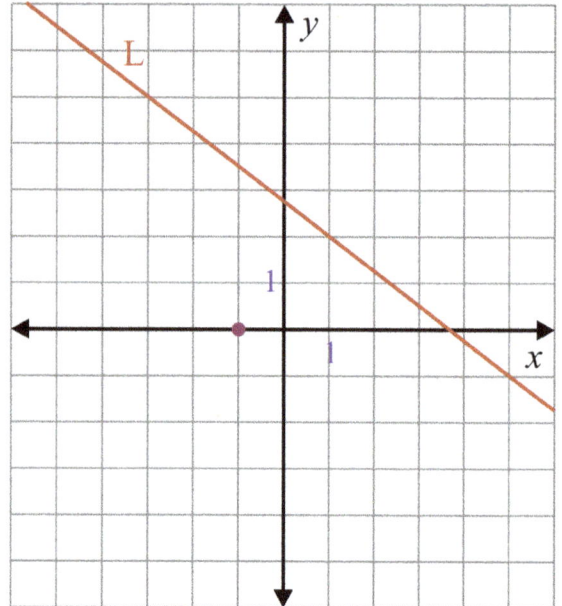

5. You see three lines. A fourth line that passes through $(0, 5)$ will be added to the picture, to form a rectangle bounded by the four lines.

   **a.** What is the slope of the fourth line?

   **b.** Plot the line.

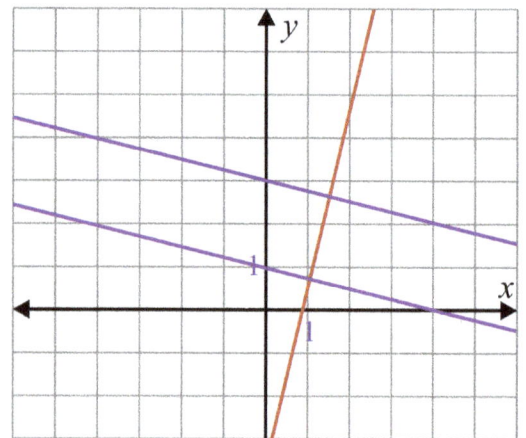

6. Are the two lines $x + 5y = 10$ and $y = 5x - 3$ perpendicular? Explain.

7. Line $t$ passes through the points A(−5, −4) and B(−1, −2). Line $s$ passes through the points A' and B' which are rotations of A and B 90° clockwise around the origin.

**a.** What is the slope of line $t$?

**b.** Graph lines $s$ and $t$.

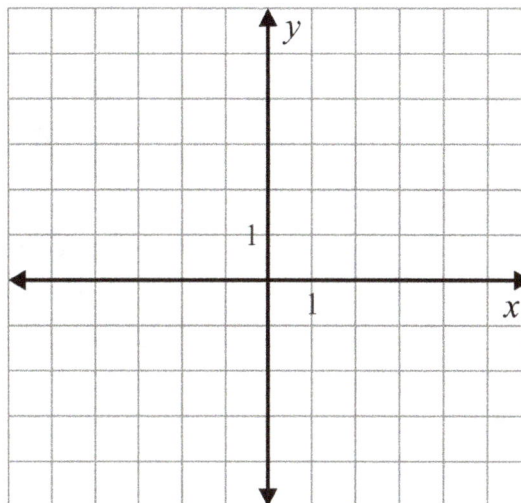

8. Find the equation of each line. Give each equation in standard form.

| | |
|---|---|
| **a.** has $y$-intercept 5 and is parallel to $y = -10x + 7$ | **b.** passes through (0, −2) and is perpendicular to $y = x$ |
| **c.** passes through (−2, 2) and is parallel to $y = 6$ | **d.** has $y$-intercept −4 and is perpendicular to $y = 2x$ |
| **e.** passes through (4, −8) and is parallel to $x = 1$ | **f.** passes through (0, 12) and is perpendicular to $y = -(1/3)x$ |

# Mixed Review Chapter 5

1. The graph shows the distance between Tyler and his home, as he goes to the grocery store and comes back. Continue graphing the function. Here is the story:

   - Tyler leaves home, driving at a constant speed of 60 km/h.

   - After five minutes, he hits a traffic jam and sits there without moving for ten minutes.

   - After that, he starts moving again at a speed of 42 km/h, and continues with that for 10 minutes.

   - Then he reaches the store, and stays there for 30 minutes.

   - On his way home, he is able to keep a steady speed of 48 km/h.

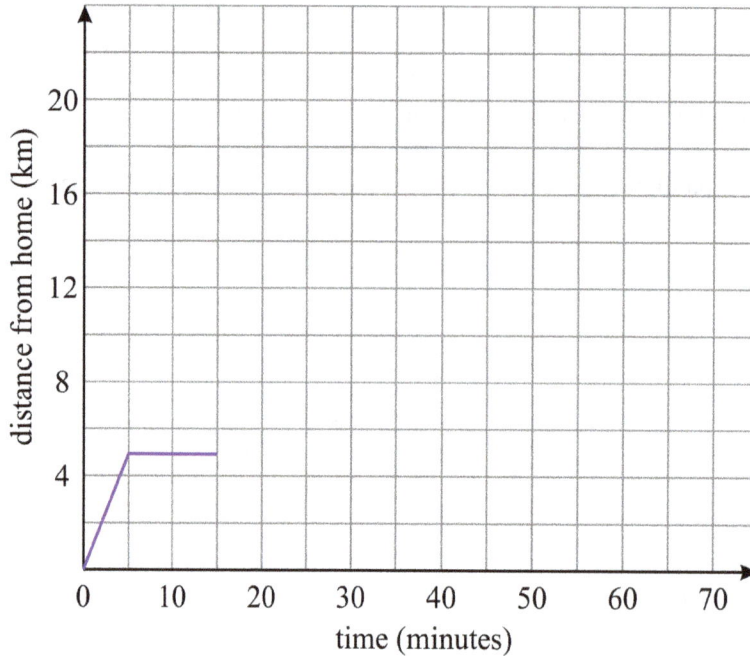

2. Prove that figure 3 is congruent to figure 4 by explaining a sequence of transformations that maps figure 3 onto figure 4.

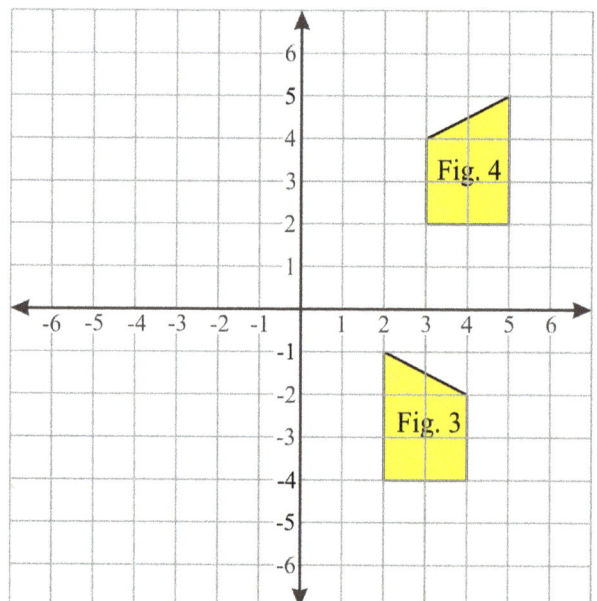

3. **a.** How many solutions does this equation have?

$$3(y - 1) = 7 + 3y$$

   **b.** Modify the equation so that it has *one* solution.

4. Two lines intersect at point C.

   **a.** Are the triangles ABC and CDE similar?
   How do you know?

   **b.** Find the value of *x*.

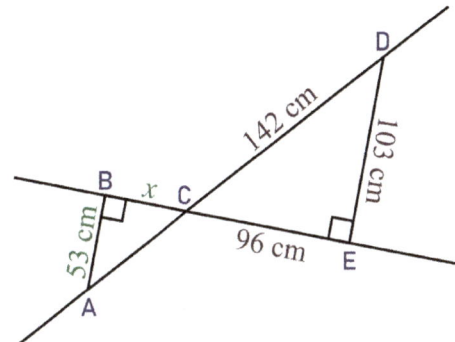

5. Which of the three functions represented below has the largest rate of change...

   **a.** in the *x*-interval [1, 3]?

   **b.** in the *x*-interval [4, 5]?

**Function 1:**

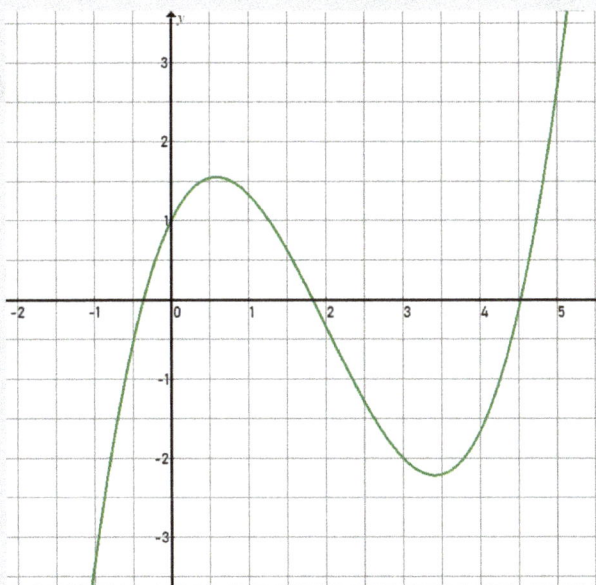

**Function 2:**

$$y = 2x - 5$$

**Function 3:**

| x | y |
|---|---|
| 1 | −4 |
| 2 | 1 |
| 3 | 6 |
| 4 | 11 |
| 5 | 16 |

6. Solve.

| a.     $10(x - 3) + 2x - 5 = 6 - 3x$ | b.     $\frac{1}{3}x - 5 = \frac{1}{4}x + 2$ |
|---|---|
| c.     $\frac{1}{6}(x - 7) = -\frac{7}{8}$ | d.     $\dfrac{5x - 2}{10} - 2 = 3x$ |

7. Mary says to Ryan, "Twenty years ago,
   I was 3/5 of your age, and now I am
   7/9 of your age." How old is Mary now?

8. Amy's piggy bank has a bunch of nickels,
   twice as many dimes, and 16 quarters.
   The total value of her coins is $12.25.
   How many of each type of coin does she have?

# Chapter 5 Review

1. Refrigerator companies make estimates of how much energy their fridges consume in typical usage. The table shows how many kilowatt-hours (kWh) of energy fridge 1 consumed over time, and the graph shows the same for fridge 2.

**Fridge 1**

| time (mo) | energy (kWh) |
|-----------|--------------|
| 2 | 75 |
| 4 | 150 |
| 6 | 225 |
| 8 | 300 |
| 10 | 375 |
| 12 | 450 |

**Fridge 2**

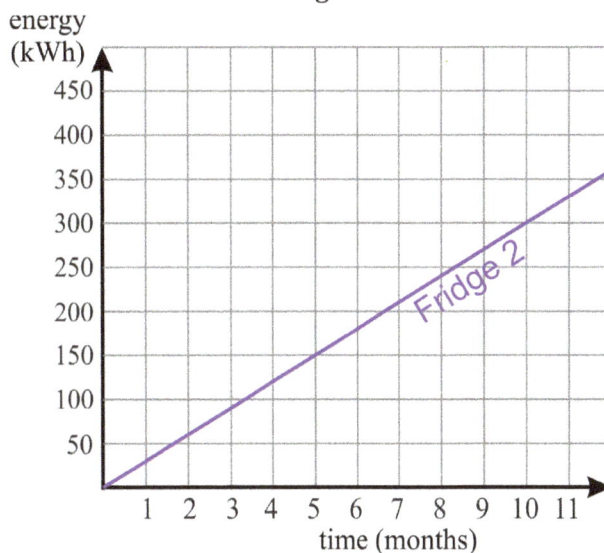

a. Which fridge consumes more electricity in a month?

   How much more?

b. Write an equation for each fridge, relating the energy (E, in kWh) and the time (t, in months).

c. Plot the equation for Fridge 1 in the grid.

d. Plot the point corresponding to the unit rate, for Fridge 1.

2. a. Find the equations of the four lines, in slope intercept form.

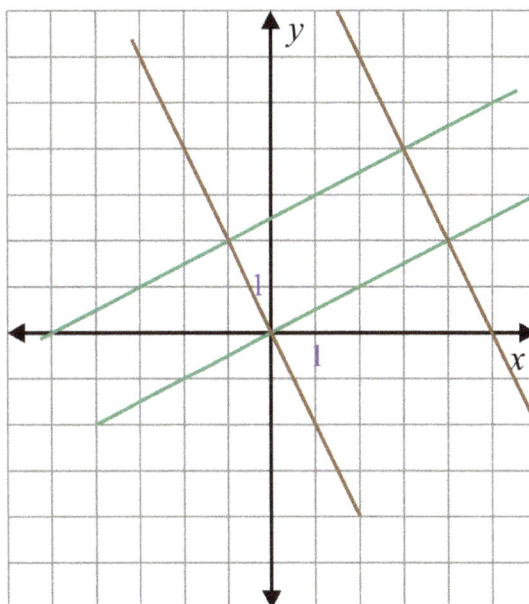

   b. (optional) Find the area of the rectangle.

3. Find the equation of each line, in slope-intercept form:

    **a.** has slope 3/4 and passes through (−2, 3)

    **b.** is horizontal and passes through (9, −10)

4. Find the slope of the lines.
   Notice the scaling.

   **a.**

   **b.**

   Now find the equations for the lines.

   **a.**

   **b.**

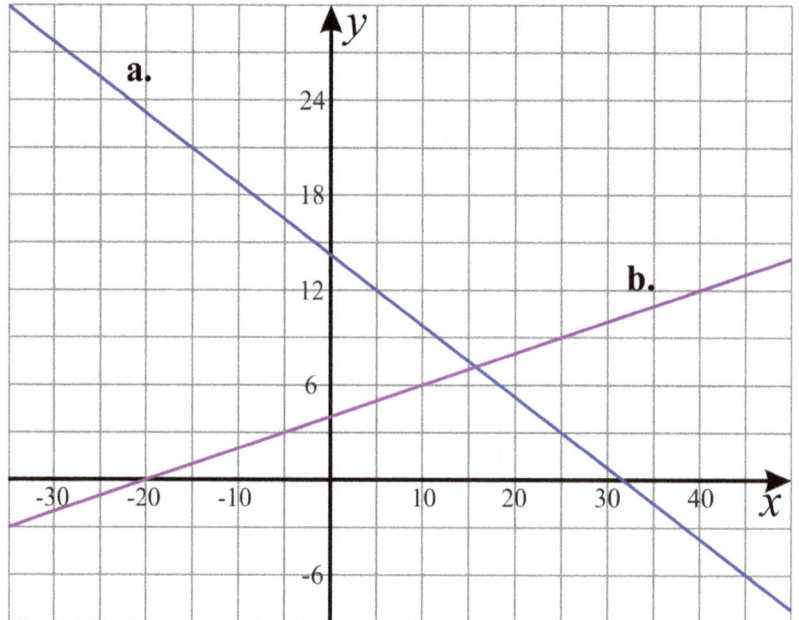

5. Do the three points fall on one line? Explain your reasoning.

   (−3, 1), (−1, −4), (1, −8)

6. Find $s$ so that the point $(s, 12)$ will fall on the same line as the points (3, 9) and (15, 18).

7. Line S passes through (−5, −2) and (0, 4). Line T is perpendicular to Line S, and passes through (1, 1).

   **a.** Find the equation of line T, in slope-intercept form.

   **b.** Write the equation also in the standard form.

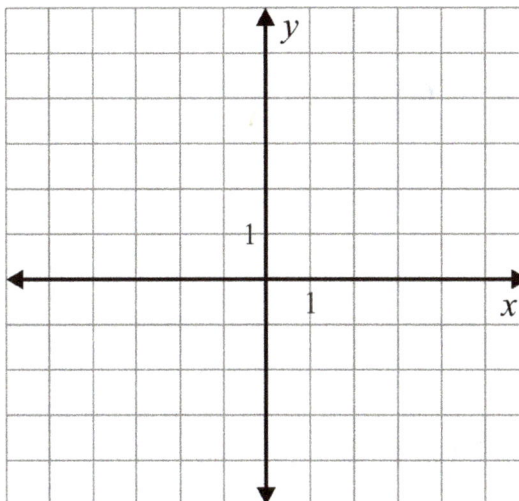

8. Mr. Henson runs a garbage pick-up business, with 12 garbage trucks. To run one truck costs him $2100 per month in maintenance costs, plus $180 a day for fuel.

   Consider the cost of running one truck as a function of time, in days (during one month only). Is this a linear relationship, a proportional relationship, or neither?

   Write an equation for it.

9. Match the descriptions and the equations.

| | |
|---|---|
| $y = (-4/3)x - 7$ | Is parallel to $x = 9$ and passes through (2, 7) |
| $3x - y = -21$ | Has $y$-intercept −4 and is perpendicular to $y = -2x$. |
| $y = -4$ | Passes through (−5, 6) and has slope 3. |
| $x - 2y = 8$ | Passes through (−9, 5) and (−3, −3) |
| $x = 2$ | Passes through (−3, 0) and (0, 9) |
| $y = 3x + 9$ | Has $y$-intercept −4 and is parallel to $y = -2$. |

10. Transform each equation of a line to the standard form, and then list its $x$ and $y$-intercepts.

| a. $y - 6 = 2(x + 2)$ | b. $-\dfrac{1}{3}x - \dfrac{3}{2}y = 1$ |
|---|---|
|  |  |

11. A heater was turned on at 10 AM in a cold, uninhabited house, to prepare it for people later that day. The graph shows the temperature of the house. The count of hours starts at 10 AM.

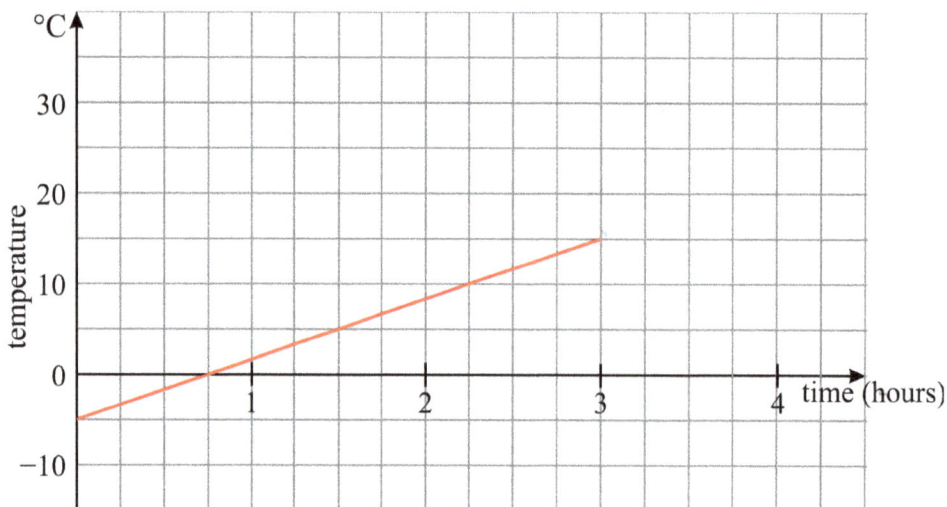

a. Write an equation for the line.

b. If the temperature continues to rise in the same fashion, what will the temperature be at 2:30 PM?

c. When will the temperature reach 22°C?

d. Let's say the heater is turned off at 1:45. What is the temperature at that time?

e. If the house had started out at a temperature of −12°C instead, and the heating process worked in the same fashion (the temperature rose at the same rate), at what time would the house reach a temperature of 22°C?

# Chapter 6: Irrational Numbers and the Pythagorean Theorem
## Introduction

We start out this chapter by studying the concept of a square root, as the opposite operation to squaring a number. In the next lesson, on irrational numbers, students find values of square roots by hand. They make a guess and then square the guess, and based on how close the square of their guess is to the radicand, they refine their guess until desired accuracy is reached. This will help solidify the concept of a square root, while also showing how most square roots are nonending decimal numbers, and how in real life, we need to use approximations of them to do calculations. Students also practise placing irrational numbers on the number line, using mental math to find their approximate location.

Next, the chapter has a review lesson on how to convert fractions to decimals. The following lesson has to do with writing decimals as fractions, and teaches a method for converting repeating decimals to fractions.

Then it is time to learn to solve simple equations that involve taking a square or cube root, over the course of two lessons. After learning to solve such equations, students are now fully ready to study the Pythagorean Theorem and apply it.

The Pythagorean Theorem is introduced in the lesson by that name. Students learn to verify that a triangle is a right triangle by checking whether it fulfils the Pythagorean Theorem. They apply their knowledge about square roots and solving equations to solve for an unknown side in a right triangle when two of the sides are given.

Next, students solve a variety of geometric and real-life problems that require the Pythagorean Theorem. This theorem is extremely important in many practical situations. Students should show their work for these word problems to include the equation that results from applying the Pythagorean Theorem to the problem and its solution.

There are literally hundreds of proofs for the Pythagorean Theorem. In this book, we present one easy proof based on geometry (not algebra). As an exercise, students are asked to supply the steps of reasoning to another geometric proof of the theorem. Students also study a proof for the converse of the theorem, which says that if the sides of a triangle fulfil the equation $a^2 + b^2 = c^2$ then the triangle is a right triangle.

Our last topic is distance between points in the coordinate grid, as this is another simple application of the Pythagorean Theorem.

## Pacing Suggestion for Chapter 6

This table does not include the chapter test as it is found in a different book (or file).
Please add one day to the pacing if you use the test.

| The Lessons in Chapter 6 | page | span | suggested pacing | your pacing |
|---|---|---|---|---|
| Square Roots ............................................................. | 65 | *4 pages* | 1 day | |
| Irrational Numbers ................................................... | 69 | *4 pages* | 1 day | |
| Cube Roots and Approximations of Irrational Numbers ... | 73 | *4 pages* | 1 day | |
| Fractions to Decimals (optional)........................................ | 77 | *(2 pages)* | (1 day) | |
| Decimals to Fractions ................................................ | 79 | *3 pages* | 1 day | |
| Square and Cube Roots as Solutions to Equations ............. | 82 | *3 pages* | 1 day | |
| More Equations that Involve Roots ................................... | 85 | *3 pages* | 1 day | |
| The Pythagorean Theorem ............................................ | 88 | *5 pages* | 2 days | |
| Applications of the Pythagorean Theorem 1 ..................... | 93 | *3 pages* | 1 day | |

## Helpful Resources on the Internet

We have compiled a list of Internet resources that match the topics in this chapter, including pages that offer:

- **online practice** for concepts;
- online **games**, or occasionally, printable games;
- **animations** and interactive **illustrations** of math concepts;
- **articles** that teach a math concept.

We heartily recommend you take a look! Many of our customers love using these resources to supplement the bookwork. You can use these resources as you see fit for extra practice, to illustrate a concept better and even just for some fun. Enjoy!

https://l.mathmammoth.com/gr8ch6

Scan me

# Square Roots

The **square** of a number is that number multiplied by itself. For example, six squared $= 6^2 = 6 \cdot 6 = 36$. (Recall that the square of 6 tells us the area of a square with sides 6 units long.)

Taking a **square root** is the opposite operation to squaring: the square root of 36 is the number that when squared, gives you 36.

There are actually two such numbers: 6 and −6. The positive one, 6, is **the principal square root** of 36. We use the "$\sqrt{\phantom{x}}$" symbol (called the "radical sign" or "radix") to signify the principal square root of a number. For example, $\sqrt{25} = 5$ because $5^2 = 25$.

The words "radish" and "radical" both come from the Latin word *radix*, meaning **root**.

Taking a square root allows us to find the side length of a square when its area is given.

Here is a way to remember what a square root is. In the picture on the right, the area of a square is written inside the square and the length of the side is written to the side:

Now, imagine the square is a radical sign that "houses" the number for the area:

**To find the (principal) square root of a number, think of a square with that area, and find the side length of that square.**

$$\boxed{49}\;7$$

$$\sqrt{\boxed{49}} = 7$$

1. Find the (principal) square roots.

| | | | |
|---|---|---|---|
| **a.** $\sqrt{100}$ | **b.** $\sqrt{64}$ | **c.** $\sqrt{4}$ | **d.** $\sqrt{0}$ |
| **e.** $\sqrt{81}$ | **f.** $\sqrt{144}$ | **g.** $\sqrt{1}$ | **h.** $\sqrt{10\,000}$ |

2. It is especially easy to find square roots of numbers that are **perfect squares**: numbers we get by squaring whole numbers. For example, 49 is a perfect square because it is $7^2$.

Fill in the list of perfect squares from $1^2$ to $20^2$ at the right:

3. Find the square roots of these perfect squares.

    **a.** $\sqrt{169}$            **b.** $\sqrt{900}$

    **c.** $\sqrt{225}$            **d.** $\sqrt{121}$

    **e.** $\sqrt{441}$            **f.** $\sqrt{8100}$

    **g.** $\sqrt{324}$            **h.** $\sqrt{400}$

    **i.** $\sqrt{6400}$           **j.** $\sqrt{25\,600}$

    **k.** $\sqrt{16\,900}$          **l.** $\sqrt{1\,000\,000}$

| $x$ | $x^2$ | $x$ | $x^2$ |
|---|---|---|---|
| 1 | 1 | 11 | ___ |
| 2 | 4 | 12 | ___ |
| 3 | 9 | 13 | ___ |
| 4 | ___ | 14 | ___ |
| ___ | ___ | 15 | ___ |
| ___ | ___ | ___ | 256 |
| ___ | 49 | ___ | 289 |
| 8 | ___ | ___ | 324 |
| 9 | ___ | ___ | ___ |
| ___ | ___ | ___ | ___ |

Most whole numbers are *not* perfect squares, and their square roots are unending decimals. (In fact, their square roots are **irrational numbers**, which means they cannot be written as a fraction, and their decimal expansions are unending decimals without any repeating patterns in the digits.)

We can handle that situation in at least three ways:

1. We can find an approximate value of such square roots **with a calculator**, rounding the answer to a reasonable accuracy. This is necessary if we're dealing with a real-life application.

2. We can find an approximate value using **guess and check**, and decimal multiplication. For example, we know that the value of $\sqrt{17}$ will be between 4 and 5 (since $\sqrt{16} = 4$ and $\sqrt{25} = 5$). We can also tell that it will be closer to 4 than 5, since 17 is very close to 16. So, we could guess that it is 4.1, square that, and based on the result, refine our guess.

3. We can **indicate such values using the square root symbol**, and not find a decimal approximation. For example, the side of a square with an area of 2 square units is $\sqrt{2}$ units. This is the preferred way in pure mathematics, and any time you want to convey an accurate value.

4. Between which two consecutive whole numbers do the following square roots lie? Do not use a calculator. Tell also which of those whole numbers the root is closer to.

   **a.** $\sqrt{5}$            **b.** $\sqrt{24}$            **c.** $\sqrt{47}$            **d.** $\sqrt{83}$

5. Tell the side of the square (exact value) when its area is given. Indicate the side length using the square root symbol, if the area is not a perfect square. Note: $u^2$ signifies square units, and $u$ signifies a unit.

| **a.** area = 25 $u^2$ | **b.** area = 1600 $u^2$ | **c.** area = 5 $u^2$ | **d.** area = 11 $u^2$ |
|---|---|---|---|
| side = _____ | side = _____ | side = _____ | side = _____ |

6. **a.** What is the area of a square, if its side measures $\sqrt{8}$ units?

   **b.** What is the value of $(\sqrt{7})^2$?

   **c.** What is the side of a square with an area of 130 square metres? Give an exact value.

---

**Example 1.** Since $0.5 \cdot 0.5 = 0.25$, then $\sqrt{0.25} = 0.5$.

**Example 2.** Since $\dfrac{2}{3} \cdot \dfrac{2}{3} = \dfrac{4}{9}$, then $\sqrt{\dfrac{4}{9}} = \dfrac{2}{3}$.

---

7. Find the square roots.

   **a.** $\sqrt{0.16}$            **b.** $\sqrt{0.01}$            **c.** $\sqrt{1.21}$

   **d.** $\sqrt{\dfrac{16}{25}}$            **e.** $\sqrt{\dfrac{100}{9}}$            **f.** $\sqrt{\dfrac{49}{36}}$

**Example 3.** The area of a square is 42.5 m². What is the side of the square?

From a calculator, $\sqrt{42.5} \approx 6.5192024052026487145829715574292$. Even this long decimal is not giving us all the decimal digits! In reality, the number would continue in an unending manner, without any patterns in the decimal digits (it is an *irrational* number).

The area was given to three significant digits, so we will do the same here, and use three significant digits in our answer. The side measures 6.52 metres.

Note: On some calculators, you first push the square root button, then the number of which you are taking the square root. On others, you first enter the number and then push the square root button. Find out which way your calculator works.

8. Find the value of these square roots with a calculator, to three decimal digits.

| a. $\sqrt{70}$ | b. $\sqrt{3}$ | c. $\sqrt{1450}$ |
|---|---|---|
| d. $\sqrt{0.45}$ | e. $\sqrt{\dfrac{5}{6}}$ | f. $\sqrt{\dfrac{31}{7}}$ |

The radical sign acts as a grouping symbol: it is as if there were brackets around the expression under the square root. In other words, $\sqrt{15+10}$ means $\sqrt{(15+10)}$.

**Example 4.** Simplify $\sqrt{5 \cdot (70+10)}$.

We simplify the expression under the square root first, and take the square root last:
$$\sqrt{5 \cdot (70+10)} = \sqrt{5 \cdot 80} = \sqrt{400} = 20$$

9. Calculate. Do not use a calculator.

| a. $\sqrt{9+16}$ | b. $\sqrt{11 \cdot 11}$ | c. $\sqrt{2 \cdot (41-9)}$ |
|---|---|---|
| d. $\sqrt{225-9^2}$ | e. $\sqrt{10^2-8^2}$ | f. $\sqrt{13^2-12^2}$ |

10. Find the value of these expressions to three decimal digits with a calculator. Note: if your calculator doesn't automatically follow the order of operations, you need to use brackets when entering the expressions. Another option is to write the intermediate results down or load them into the calculator's memory.

| a. $\sqrt{5.6^2-2.1^2}$ | b. $\sqrt{45.7^2+38.12^2}$ |
|---|---|

11. **a.** What is the area of a square if its side measures $\sqrt{1600}$ cm?

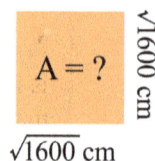

A = ?

$\sqrt{1600}$ cm

$\sqrt{1600}$ cm

    **b.** What is the area of a square if its side measures $\sqrt{37}$ m?

12. **a.** Sketch a square with an area of 18 square centimetres.

    **b.** What is its perimeter, to two decimal digits?

13. **a.** Sketch a square with a perimeter of 18 cm.

    **b.** What is its area, to two decimal digits?

14. Place the letter of each expression in the box below its value, solving the riddle.

| | | | |
|---|---|---|---|
| **R** $\sqrt{4.41}$ | **O** $\sqrt{1.2 \cdot 0.3}$ | **E** $\sqrt{0.16}$ | **H** $\sqrt{169}$ |
| **G** $\sqrt{100}$ | **T** $\sqrt{\dfrac{225}{9}}$ | **Q** $\sqrt{6^2 - 1}$ | **R** $\sqrt{10^2 - 6^2}$ |
| **S** $\sqrt{\dfrac{25}{4}}$ | **E** $\sqrt{\dfrac{64}{25}}$ | **U** $\sqrt{900}$ | **S** $\sqrt{\dfrac{49}{100}}$ |
| **V** $\sqrt{144}$ | **E** $\sqrt{1.21}$ | **A** $\sqrt{100(20 + 5)}$ | **I** $\sqrt{41 \cdot 41}$ |
| **O** $\sqrt{43 + 51}$ | **M** $\sqrt{0.0016}$ | **S** $\sqrt{10\ 000}$ | **T** $\sqrt{4(20 - 8)}$ |

<u>*Why do plants hate math?*</u>

Because it...

| 10 | 41 | 12 | 0.4 | 2.5 | | 5 | 13 | 1.1 | 0.04 | | 100 | $\sqrt{35}$ | 30 | 50 | 2.1 | 1 3/5 | | 8 | $\sqrt{94}$ | 0.6 | $\sqrt{48}$ | 0.7 |
|---|---|---|---|---|---|---|---|---|---|---|---|---|---|---|---|---|---|---|---|---|---|---|
| | | | | | | | | | | | | | | | | | | | | | | |

**Puzzle Corner**

Make number 19 on a broken calculator that only has these buttons:

| 5 | 6 | + | × | √ | ( | ) |

# Irrational Numbers

The square roots of perfect squares are whole numbers. However, most numbers, such as 2, 5, and 17, are not perfect squares. We can find a **decimal approximation** to these types of square roots, either with a calculator, or manually, by using the techniques of squaring and guess-and-check.

**Example 1.** Find the value of $\sqrt{19}$ to two decimal digits, without using a calculator's square root function.

First we find two consecutive perfect squares so that 19 is between them: **16** < 19 < **25**. From that we know that $4 < \sqrt{19} < 5$. Also, since 19 is closer to 16 than to 25, we expect $\sqrt{19}$ to be closer to 4 than to 5.

So let's choose 4.3 and 4.4 as our initial guesses for the value of $\sqrt{19}$, square the guesses, and check how close to 19 we get.

| Low Guess | $(LG)^2$ | $(HG)^2$ | High Guess |
|---|---|---|---|
| 4.3 | 18.49 | 19.36 | 4.4 |

*Note:* $(LG)^2$ means low guess squared, and $(HG)^2$ means high guess squared.

From the table above, we can see that $\sqrt{19}$ is indeed between 4.3 and 4.4, and that it is probably <u>closer to 4.4</u> than it is to 4.3 (because 19.36 is closer to 19 than 18.49 is). Let's try 4.36 and 4.37 next.

| Low Guess | $(LG)^2$ | $(HG)^2$ | High Guess |
|---|---|---|---|
| 4.3 | 18.49 | 19.36 | 4.4 |
| 4.36 | 19.0096 | 19.0969 | 4.37 |

Oops! $\sqrt{19}$ is not between 4.36 and 4.37. Both of those are too high. Let's try 4.35 and 4.36 next.

| Low Guess | $(LG)^2$ | $(HG)^2$ | High Guess |
|---|---|---|---|
| 4.3 | 18.49 | 19.36 | 4.4 |
| 4.35 | 18.9225 | 19.0096 | 4.36 |

Now we know that $\sqrt{19}$ is between 4.35 and 4.36 and closer to 4.36 than it is to 4.35 (because 19.0096 is much closer to 19 than 18.9225 is). This means that **to two decimal digits, $\sqrt{19} = 4.36$.**

1. Continue refining the decimal approximation to $\sqrt{19}$, to three decimal digits. You may use a calculator to multiply, but do not use its square root function.

| Low Guess | $(LG)^2$ | $(HG)^2$ | High Guess |
|---|---|---|---|
| 4.35 | 18.9225 | 19.0096 | 4.36 |
| | | | |
| | | | |
| | | | |

2. Use only multiplication (squaring) to guess and check the values of the following square roots to two decimal digits. You may use a calculator, but not the square root function of the calculator.

   **a.** $\sqrt{7}$

   | Low Guess | (LG)² | (HG)² | High Guess |
   |---|---|---|---|
   |  |  |  |  |
   |  |  |  |  |
   |  |  |  |  |
   |  |  |  |  |

   **b.** $\sqrt{51}$

   | Low Guess | (LG)² | (HG)² | High Guess |
   |---|---|---|---|
   |  |  |  |  |
   |  |  |  |  |
   |  |  |  |  |
   |  |  |  |  |

   **c.** $\sqrt{99}$

   | Low Guess | (LG)² | (HG)² | High Guess |
   |---|---|---|---|
   |  |  |  |  |
   |  |  |  |  |
   |  |  |  |  |
   |  |  |  |  |

3. Sarah claims that $\sqrt{11}$ equals exactly 3.317. Is she correct? Explain.

4. Prove that $\sqrt{2}$ cannot equal the fraction $\dfrac{71}{50}$.

Recall that **a rational number** is a number that **can be written as a ratio** (or as a fraction) **of two integers**, with a nonzero denominator. For example, 14/13 is a rational number, and so are −6 (= −6/1) and 7.89 (= 789/100).

An **irrational number** is just the opposite: one that *cannot* **be written as a ratio of two integers**.

Each rational number, being a fraction, can also be written as a decimal. The decimal form of a rational number <u>either ends</u>, or is <u>unending with a repeating pattern</u> in its digits.

The decimal expansion of an irrational number is <u>unending and has no repeating pattern in the decimal digits</u>.

---

**Examples.**

- The rational number 2/5 as a decimal is 0.4 — a decimal that ends.
- The fraction 1/3 = 0.3333... with 3 repeating has an unending decimal expansion. We can also write 1/3 = 0.$\overline{3}$ where the line over the digit 3 means that that digit repeats indefinitely.
- 1.4$\overline{08}$ means 1.408080808... with "08" repeating. Since the decimal expansion repeats, this is a rational number, and thus can be written as a fraction: it is actually 1394/990.
- The number $\pi$ is the most famous irrational number. Its decimal expansion starts out like this: 3.14159265358979323846264338327795..., and never has any repeating pattern.
- Another famous irrational number is the Golden Ratio: $(1 + \sqrt{5})/2$, with an approximate value of 1.618.
- If a whole number is not a perfect square, its square root is an irrational number. So, $\sqrt{28}$ , $\sqrt{97}$, and $\sqrt{45}$ all are irrational numbers. And there are many others!

---

**Example 2.** Determine whether the numbers $\dfrac{\sqrt{5}}{2}$ and 0.989898 are rational or irrational.

The first, $\sqrt{5}/2$, is irrational. This is because $\sqrt{5}$ is irrational, and if you divide an irrational number by a rational number, you cannot get a rational number. (See the answer key for Puzzle Corner for proof.)

The same is true for other operations: if $x$ is irrational and $r$ is rational, then all of these are irrational: $x + r$, $x − r$, $xr$, and $x/r$. It follows that, for example, $7\pi$ is irrational.

The second, 0.989898, is a decimal that ends, so it is rational. (As a fraction, it is 989898/1000000.)

---

5. Determine whether the following numbers are rational or irrational, and explain why.

**a.** 0.928

**b.** $\sqrt{128}$

**c.** 0.55$\overline{812}$

**d.** 6.050606

**e.** $\pi/2$

**f.** $\sqrt{100}$

**g.** 0.20$\overline{8}$

**h.** $5\sqrt{3}$

**i.** $\dfrac{\sqrt{15}}{3}$

**j.** $\sqrt{\dfrac{25}{4}}$

6. This is Henry's work. Some of it is in error — but he can learn! Correct the ones that are wrong, and explain why.

   **a.** 1.272727 is rational because its decimal expansion repeats.

   **b.** $3\sqrt{49}$ is irrational because it has a square root.

   **c.** $\dfrac{\pi}{3}$ is rational because it is a fraction.

   **d.** $\dfrac{15}{4}$ is rational because it is a fraction.

7. Place the numbers in the correct places in the diagram of *real numbers* = the set of both rational and irrational numbers. Note: the set of whole numbers is {0, 1, 2, 3, 4, 5, ...}.

$$6,\ -76,\ \frac{22}{7},\ \sqrt{81},\ \sqrt{8},\ 4.05,\ \frac{\pi}{7},\ \sqrt{2},\ 0.2\overline{7},\ \frac{\sqrt{21}}{8},\ 0,\ -\frac{1}{8},\ \sqrt{7}+1,\ -\frac{45}{9},\ 3\sqrt{3},\ \frac{5}{\sqrt{100}},\ \sqrt{900}$$

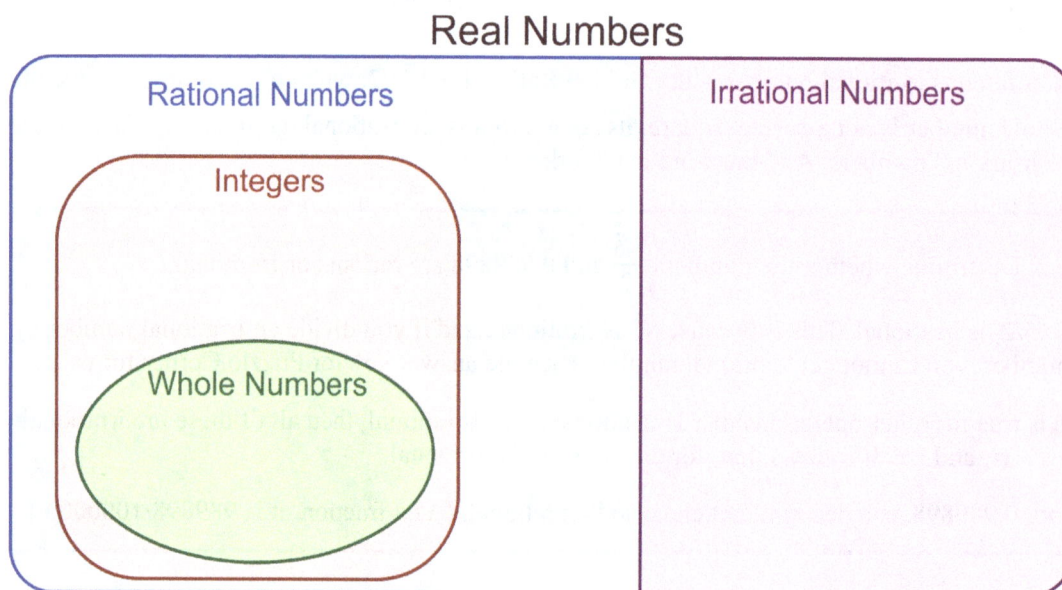

## Real Numbers

| Rational Numbers | Irrational Numbers |
|---|---|
| Integers | |
| Whole Numbers | |

**a.** Read the proof that if $x$ is irrational and $r$ is rational, then $x + r$ is irrational.

<u>Proof.</u> Suppose the contrary, that $x + r = s$ where $s$ is rational. Then, $x = s - r$. Since $s$ and $r$ can be written as fractions, their difference, $s - r$, is also a fraction, or a rational number.

This would mean $x$ is rational, which is a contradiction. Therefore our original supposition cannot be correct. Thus, $x + r$ must be irrational.

**b.** Prove that an irrational number divided by a rational number is irrational.

**Puzzle Corner**

# Cube Roots and Approximations of Irrational Numbers

Similarly to the square root, we can take a **cube root** of a number.

Recall that the **cube of a number** is that number multiplied by itself three times. For example, two cubed = $2^3 = 2 \cdot 2 \cdot 2 = 8$. This gives us the volume of a cube with edges 2 units long.

The cube root of 8 is 2. We write it as $\sqrt[3]{8} = 2$. Notice the little "3" that is added to the radical sign to signify a cube root.

**Example 1.** Since $(-3)(-3)(-3) = -27$, then $\sqrt[3]{-27} = -3$.

Like square roots, most cube roots are irrational numbers. When it comes to integers, only the cube roots of perfect cubes are rational; the rest are irrational.

1. Find the cube roots without a calculator.

| a. $\sqrt[3]{27}$ | b. $\sqrt[3]{125}$ | c. $\sqrt[3]{64}$ | d. $\sqrt[3]{1000}$ |
|---|---|---|---|
| e. $\sqrt[3]{1}$ | f. $\sqrt[3]{216}$ | g. $\sqrt[3]{27\,000}$ | h. $\sqrt[3]{-8}$ |
| i. $\sqrt[3]{-1}$ | j. $\sqrt[3]{-125}$ | k. $\sqrt[3]{0}$ | l. $\sqrt[3]{-8000}$ |

2. **a.** The volume of a cube is 216 cm$^3$. How long is its edge?

   **b.** What is $(\sqrt[3]{4})^3$?

   **c.** If the edge of a cube measures 50 cm, find its volume.

   **d.** If the volume of a cube is 729 cm$^3$, find its surface area.

3. (optional) Find the cube roots of these fractions and decimals, without a calculator.

| a. $\sqrt[3]{0.008}$ | b. $\sqrt[3]{0.125}$ | c. $\sqrt[3]{-0.027}$ |
|---|---|---|
| d. $\sqrt[3]{\dfrac{8}{125}}$ | e. $\sqrt[3]{\dfrac{64}{27}}$ | f. $\sqrt[3]{-\dfrac{1}{8}}$ |

**Example 2.** We can know that $\sqrt{98}$ lies between 9 and 10, because $9 = \sqrt{81} < \sqrt{98} < \sqrt{100} = 10$.

We can even tell it is much closer to 10 than to 9, since 98 is much closer to 100 than to 81.

From that, we can estimate that $2\sqrt{98}$ is slightly less than 20, and that $\sqrt{98} + 4$ is slightly less than 24.

**Example 3.** The opposite of $\sqrt{2}$ is $-\sqrt{2}$. Since $\sqrt{2}$ is approximately 1.41, then $-\sqrt{2} \approx -1.41$.

4. Find between which two whole numbers the root lies. Notice some of them are cube roots.

| | | |
|---|---|---|
| **a.** _____ < $\sqrt{31}$ < _____ | **b.** _____ < $\sqrt{65}$ < _____ | **c.** _____ < $\sqrt{87}$ < _____ |
| **d.** _____ < $-\sqrt{5}$ < _____ | **e.** _____ < $-\sqrt{44}$ < _____ | **f.** _____ < $-\sqrt{50}$ < _____ |
| **g.** _____ < $\sqrt[3]{7}$ < _____ | **h.** _____ < $\sqrt[3]{37}$ < _____ | **i.** _____ < $\sqrt[3]{101}$ < _____ |

5. Plot the following numbers *approximately* on the number line. Do not use a calculator, but think about between which two integers the root lies, and whether it is close to one of those integers.

$\sqrt{15}$　　　$\sqrt{47}/2$　　　$\sqrt[3]{9}$　　　$-\sqrt[3]{27}$　　　$-\sqrt{10}$　　　$\sqrt{66}/2$　　　$\pi$　　$\sqrt{18} + 1$

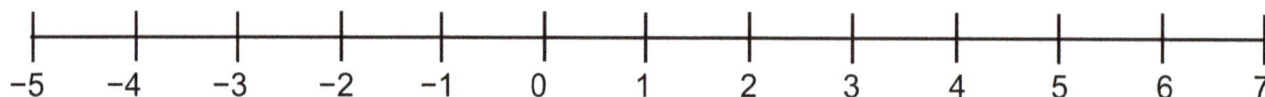

6. Compare, writing >, <, or = between the numbers. Think between which two whole numbers the root lies, using mental math.

| | | | |
|---|---|---|---|
| **a.** 5 ☐ $\sqrt{27}$ | **b.** $\sqrt{48}$ ☐ 7 | **c.** $\sqrt{18}$ ☐ 4 | **d.** $\sqrt[3]{9}$ ☐ 2 |
| **e.** 2 ☐ $\sqrt{2} + 1$ | **f.** $\sqrt{32} + 1$ ☐ 6 | **g.** $\sqrt{43} + 5$ ☐ 10 | **h.** $\sqrt{88} - 3$ ☐ 7 |

7. **a.** Between which two whole numbers does $\sqrt{30}$ lie? And $\sqrt{60}$?

**b.** Use your answers to (a) to determine whether $2\sqrt{30}$ is equal to $\sqrt{2 \cdot 30}$.

8. Is $\dfrac{\sqrt{50}}{2}$ equal to $\sqrt{\dfrac{50}{2}}$ ? Explain your reasoning.

9. Use the decimal approximations of common irrational numbers on the right to estimate the value of the expressions below, to one decimal digit. Use mental math and paper-and-pencil calculations, not a calculator.

$$\pi \approx 3.14$$

$$\sqrt{2} \approx 1.41$$

$$\sqrt{5} \approx 2.24$$

a. $5\sqrt{2}$

b. $\pi^2$

c. $\sqrt{5} - \sqrt{2}$

d. $2\sqrt{5} - 5\sqrt{2}$

10. a. Find an approximation to $\sqrt{11}$ to one decimal digit, without using the square root function of a calculator.

b. Use the approximation you found to estimate the values of $\sqrt{11} - \sqrt{2}$ and $3\sqrt{11}$.

11. Sarah has used the method of squaring her guesses to find out that $\sqrt{45}$ is between 6.7 and 6.8. How can she continue from this point to get a better approximation? Do it for her, to two decimal digits.

*Use these exercises for additional practice.*

12. Order the numbers from smallest to greatest. Estimate the value of the roots, thinking between which two whole numbers each square root lies, using mental math.

$\sqrt{5} - 1$  $\qquad$ $\sqrt[3]{1}$ $\qquad$ $\sqrt{19}/2$ $\qquad$ $\sqrt[3]{100}$ $\qquad$ $\sqrt[3]{8}$ $\qquad$ $\sqrt{13}$ $\qquad$ $\sqrt{9}$ $\qquad$ $2\pi$ $\qquad$ $\sqrt{22} + 1$

13. Plot the following numbers *approximately* on the number line. Do not use a calculator, but think about between which two integers the root lies, and whether it is close to one of those integers.

   **a.** $-2\sqrt{2}$ $\qquad\qquad$ **b.** $\sqrt{80}/3$ $\qquad\qquad$ **c.** $\sqrt{27} - 1$ $\qquad\qquad$ **d.** $-\sqrt{5} + 7$

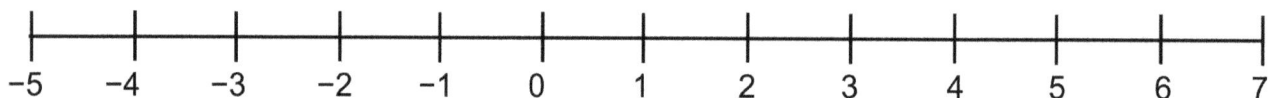

14. Plot the following numbers *approximately* on the number line.

   **a.** $-\sqrt{2} - 3$ $\qquad$ **b.** $-\pi$ $\qquad$ **c.** $-\sqrt[3]{9}$ $\qquad$ **d.** $-\sqrt{36} + 9$ $\qquad$ **e.** $-\sqrt{26}/2$

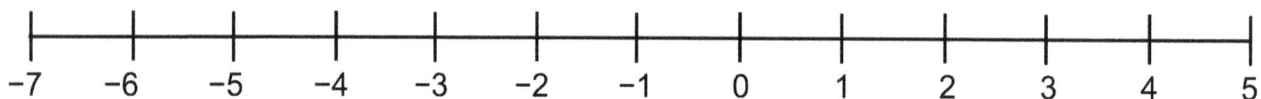

# Fractions to Decimals
## (This lesson is review, and optional.)

Each fraction is a rational number (by definition!). Each fraction can be written as a decimal. It will either be a terminating decimal, or a non-terminating repeating decimal.

It is easy to rewrite a fraction as a decimal when the denominator is a power of ten. However, when it is not (which is most of the time), simply treat the fraction as a division and divide. You will get either a **terminating decimal** or a non-terminating **repeating decimal**. See the examples below.

**1. The denominator is a power of ten** or the fraction can be simplified so that it is. In this case, writing the fraction as a decimal is straightforward. Simply write out the numerator. Then add the decimal point based on the fact that the number of zeros in the power of ten tells you the number of decimal digits.

**Examples 1.**  $\dfrac{7809}{100} = 78.09$    $\dfrac{1458}{1000} = 1.458$    $\dfrac{506}{100\,000} = 0.00506$    $\dfrac{33}{30} = \dfrac{11}{10} = 1.1$

**2. The denominator is a factor of a power of ten.** Convert the fraction into one with a denominator that is a power of ten. Then do as in case (1) above.

**Examples 2.**  $\dfrac{9}{20} = \dfrac{45}{100} = 0.45$    $\dfrac{2}{125} = \dfrac{16}{1000} = 0.016$    $\dfrac{9}{8} = \dfrac{1125}{1000} = 1.125$

**3. Use division** (long division or with a calculator). This method works in all cases, even if the denominator happens to be a power of ten or a factor of a power of ten.

**Example 3.** Write $\dfrac{31}{40}$ as a decimal.

This division terminates (comes out even) after just three decimal digits.

We get $\dfrac{31}{40} = 0.775$. This is a **terminating decimal**.

(The fact the division was even means that the denominator 40 is a factor of some power of ten, and so we could have used method 2 from above. In this case, $1000 = 40 \cdot 25$.)

```
         0 0.7 7 5
    40) 3 1.0 0 0 0
        -2 8 0
          3 0 0
        -  2 8 0
            2 0 0
          - 2 0 0
                0
```

**Example 4.** Write $\dfrac{18}{11}$ as a decimal.

We write 18 as 18.0000 in the long division "corner" and divide by 11. Notice how the digits "63" in the quotient, and the remainders 40 and 70, start repeating.

So $\dfrac{18}{11} = 1.\overline{63}$.

The fraction 18/11 equals $1.\overline{63}$, which is a **repeating decimal**.

```
         0 1.6 3 6 3
    11) 1 8.0 0 0 0
       -1 1
          7 0
        - 6 6
           4 0
         - 3 3
            7 0
          - 6 6
             4 0
           - 3 3
              7
```

1. Write the fractions as decimals.

| a. $\dfrac{2}{100} =$ | b. $\dfrac{278}{10\,000} =$ | c. $\dfrac{55\,073}{1\,000\,000} =$ |
|---|---|---|
| d. $\dfrac{4508}{1000} =$ | e. $\dfrac{56\,330}{100} =$ | f. $\dfrac{40\,309}{10\,000} =$ |

2. Write the fractions as decimals.

| a. $\dfrac{2}{5} =$ | b. $\dfrac{24}{25} =$ | c. $\dfrac{54}{200} =$ |
|---|---|---|
| d. $\dfrac{7}{4} =$ | e. $\dfrac{330}{250} =$ | f. $\dfrac{7}{125} =$ |

3. Write as decimals. Use long division, and calculate each answer to at least six decimal places. If you find a repeating pattern, give the repeating part. If you don't, round your answer to five decimals.

| a. $\dfrac{5}{9}$ | b. $\dfrac{508}{27}$ | c. $\dfrac{23}{61}$ |
|---|---|---|
|  |  |  |

# Decimals to Fractions

**When a decimal terminates**, it is straightforward to write it as a fraction:

1. Copy all the digits (without a decimal point and the leading zeros) to be the numerator.
2. You will find the denominator by checking the number of decimal places: for two decimal places, the denominator is 100, for four decimal places, the denominator is 10 000, and for $n$ decimal places, it is $10^n$.

**Example 1.**   $0.00507 = \dfrac{507}{100\,000}$     $5.256 = \dfrac{5256}{1000}$     $0.0818 = \dfrac{818}{10\,000}$

**When a decimal does not terminate and has a repeating pattern in the decimal digits, it is a rational number**, and we can write it as a fraction. The examples show you how.

**Example 2.** To write $0.2\overline{38}$ as a fraction, we multiply it by some power of ten in such a manner that when we subtract this multiple of $0.2\overline{38}$ and $0.2\overline{38}$, the repeating digits are eliminated in the subtraction.

Since there are *two* digits that repeat, we multiply $0.2\overline{38}$ by $10^2$ or 100. The digits will repeat in $100 \cdot 0.2\overline{38}$ in the same places that they do in $0.2\overline{38}$. This means that when we subtract $100 \cdot 0.2\overline{38} - 0.2\overline{38}$, the repeating digits will be eliminated.

Let $x = 0.2\overline{38}$. We will calculate $100x$ and then subtract $100x$ and $x$. See the equations on the right.

Remember to line up the decimal points carefully, so that the digits that repeat will be in the same places.

We subtract the left sides of the equations, and also the right sides. This leaves $99x$ on the left side, and 23.6 on the right side. From this new equation $99x = 23.6$, we can solve $x$: it is 23.6/99.

$$
\begin{aligned}
100x &= 23.\,8383838\ldots \\
-\quad x &= \;\;0.\,238383838\ldots \\
\hline
99x &= 23.\,6
\end{aligned}
$$

$$x = 23.6/99 = 236/990$$

$$= 118/495$$

Lastly, we multiply both the numerator and the denominator of that expression by 10, to get 236/990. This can still be simplified to **118/495**.

**Example 3.** Write $0.41\overline{509}$ as a fraction.

Let $y = 0.41\overline{509}$. Since there are three repeating digits, we will multiply $y$ by $10^3$ or 1000, and then subtract $1000y$ and $y$.

See the process on the right. We get $y = \dfrac{41\,468}{99\,900}$ .

$$
\begin{aligned}
1000y &= 415.09\,509509\ldots \\
-\quad y &= \;\;\;\;0.41\,509509509\ldots \\
\hline
999y &= 414.68
\end{aligned}
$$

$$y = 414.68/999 = 41\,468/99\,900$$

(Now, this *can* be simplified to 10 367/24 975 but since factorization takes time, you do not have to simplify the fractions in the exercises when the numerators and the denominators are large.)

1. Write as a fraction. Simplify if you can, but it is not required.

| a. 93.82 | b. 0.333 | c. 2.05056 |
|----------|----------|------------|
| d. 61.098 | e. 0.0000045 | f. 4.932048 |

2. Are the decimals $0.\overline{2}$ and 0.222 the same? If not, what is their difference?

3. Write each repeating decimal as a fraction.

| a. $0.\overline{4}$ | b. $0.2\overline{1}$ |
|---------------------|----------------------|
| c. $0.\overline{954}$ | d. $2.5\overline{32}$ |

4. Match the fractions and the decimals.

0.666   $0.\overline{51}$   $0.\overline{6}$   $0.0\overline{6}$   0.51   0.051   0.066   $0.0\overline{51}$

$\dfrac{51}{100}$   $\dfrac{2}{3}$   $\dfrac{66}{1000}$   $\dfrac{66}{990}$   $\dfrac{51}{99}$   $\dfrac{51}{990}$   $\dfrac{666}{1000}$   $\dfrac{51}{1000}$

5. Erica and Eric were comparing their methods and results for writing the decimal $2.1\overline{7}$ as a fraction.

Erica said, "Since there is one repeating digit, I multiplied it by 10, to get $21.\overline{7}$, and then subtracted $10x - x$ to get $9x = 19.6$. So, I got $x = 19.6/9 = 196/90 = 98/45$."

Eric said, "I multiplied it by 100, and got $100x - x = 217.\overline{7} - 2.1\overline{7} = 215.6$. From that, $x = 215.6/99 = 2156/990$."

Who is correct?

6. Janet's calculator gave her the result that $305/55494 = 0.0054960896673514253793202868786$. Janet said that since there is no repeating pattern here, that this is an irrational number. Is she correct?

7. Write each repeating decimal as a fraction.

| a. $0.\overline{256}$ | b. $3.05\overline{94}$ |
|---|---|
| c. $0.03\overline{2199}$ | d. $1.\overline{36309}$ |

# Square and Cube Roots as Solutions to Equations

**Example 1.** Solve $x^2 = 81$.

We can use mental math: one obvious solution is $x = 9$. However, there is also another solution!
It is not only true that $9^2 = 81$, but $(-9)^2 = 81$ also, so $x = -9$ is a second solution to this equation.

**Example 2.** Solve $x^2 = 48$.

This time, we cannot solve the equation with mental math, but we will *take a square root of both sides of the equation*. This will undo the squaring, and isolate $x$, because taking a square root and squaring are opposite operations.

$$x^2 = 48$$

$\sqrt{\phantom{-}}$   The radicand symbol signifies taking a square root of both sides of the equation.

$$x = \sqrt{48} \approx 6.93$$
$$\text{or } x = -\sqrt{48} \approx -6.93$$

Since taking a square root undoes the squaring, $x$ is now left alone on the left side. Notice that there are two solutions: the square root of 48 and the negative square root of 48.

Notice that $-\sqrt{48}$ doesn't mean that we take a square root of a negative number. Instead, $-\sqrt{48}$ means we *first* take the square root of 48 (a positive number) and then take the opposite of that result.

1. Solve. Remember, there will be two solutions: one positive and one negative. When the two answers aren't integers, give them as square roots and also as decimals rounded to two decimal digits.

| | |
|---|---|
| **a.**     $x^2 = 25$ | **b.**     $y^2 = 3600$ |
| **c.**     $x^2 = 500$ | **d.**     $z^2 = 11$ |
| **e.**     $w^2 = 287$ | **f.**     $q^2 = 1\ 000\ 000$ |

The situation is similar with equations where the variable is cubed.

**Example 3.** Solve $x^3 = 125$.

Since 125 is a perfect cube, the solution is easy to find with mental math.

$$x^3 = 125 \quad \Big|\ \sqrt[3]{\ }$$
$$x = \sqrt[3]{125} = 5$$

With cube roots, **there is no other solution.** For example, in this case, $(-5)^3$ does not equal 125.

**Example 4.** Solve $x^3 = 35$.

We will take the cube root of both sides of the equation. This will undo the cubing and isolate $x$.

$$x^3 = 35 \quad \Big|\ \sqrt[3]{\ }$$
$$x = \sqrt[3]{35} \approx 3.27$$

Since 35 is not a perfect cube, $\sqrt[3]{35}$ is an irrational number. Depending on context, we might give the answer in root form, or with a decimal approximation.

Hint: If your calculator does not have a button for the cube root, you can instead use the button for exponentiation, with the exponent 1/3. For example, on my computer calculator, I enter $\sqrt[3]{23}$ this way:

| 2 | 3 | $x^y$ | ( | 1 | / | 3 | ) |

2. Solve. If the root is not a whole number, give it rounded to two decimal digits.

| a. | b. | c. |
|---|---|---|
| $x^3 = 64$ | $n^3 = 216$ | $z^3 = 27\,000$ |
| d. | e. | f. |
| $x^3 = 7$ | $b^3 = 109$ | $a^3 = 18$ |

3. Below, the variable V signifies the volume of a cube. Find the edge of the cube ($s$) when the volume is given. Give the answer to the same amount of significant digits as the given volume.

| a. | b. | c. |
|---|---|---|
| $V = 510 \text{ m}^3$ | $V = 24\,500 \text{ cm}^3$ | $V = 5.83 \text{ m}^3$ |
| $s =$ | | |

4. Find the surface area of a cube with a volume of $14.2 \text{ m}^3$.

83

5. Solve. Since these are pure mathematical problems, give the solutions in root form. Check your solutions.

| | |
|---|---|
| **a.** $a^2 - 8 = 37$ | **b.** $y^2 + 100 = 1000$ |
| **c.** $b^2 + 1.5 = 6.4$ | **d.** $x^2 - 26 = 709$ |

6. Solve. Give the solutions in exact form. Check your solutions.

| | |
|---|---|
| **a.** $x^3 - 5 = 59$ | **b.** $x^3 + 78 = 437$ |

# More Equations that Involve Roots

**Example 1.** $3x^2 = 40$ $\quad\div 3$

Again, we want to isolate the term $x^2$, so we first divide both sides by 3.

$x^2 = 40/3$ $\quad\sqrt{\phantom{x}}$

This symbol signifies taking a square root of both sides of the equation.

$x = \sqrt{40/3}$ or $x = -\sqrt{40/3}$

There are two solutions, as usual.

$x \approx 3.651$ or $x \approx -3.651$

These are the decimal approximations.

Here is a check using the rounded positive root:

$3 \cdot 3.651^2 \overset{?}{=} 40$

$3 \cdot 13.329801 \overset{?}{=} 40$

$39.989403 \approx 40$ ✓

Here is a check using the exact positive root:

$3 \cdot (\sqrt{40/3})^2 \overset{?}{=} 40$

$3 \cdot 40/3 \overset{?}{=} 40$

$40 = 40$ ✓

1. Solve. Give the solutions both in exact format and as decimals rounded to three decimals.

| | |
|---|---|
| **a.** $\quad 5x^2 = 125$ | **b.** $\quad 8.2b^2 = 319$ |
| **c.** $\quad a^2 + 4.5 = 10.7$ | **d.** $\quad 12b^2 = 36\,000$ |

**Example 2.** $x^2 + 7^2 = 12^2$      First simplify.

$x^2 + 49 = 144$      Now it looks more familiar. Subtract 49.

$x^2 = 95$      Now we take a square root of both sides.

$x = \sqrt{95}$ or $x = -\sqrt{95}$      There are two solutions, as usual.

Check:

$(\sqrt{95})^2 + 7^2 \overset{?}{=} 12^2$

$95 + 49 \overset{?}{=} 144$

$144 = 144$ ✔

2. Solve. Give the solutions in exact form. Check your solutions.

**a.**      $a^2 + 3^2 = 7^2$

**b.**      $43^2 + x^2 = 51^2$

3. Solve. Give the solutions rounded to two decimals. Check your solutions.

**a.**      $s^2 = 2.1^2 + 5.4^2$

**b.**      $21^2 - w^2 = 15^2$

**c.**      $121^2 - x^2 = 56$

**d.**      $a^2 - 4.5^2 = 5.78$

4. Solve equations involving cube roots, also. Give each solution rounded to three decimals.

| a. $56 + s^3 = 542$ | b. $5x^3 = 180$ | c. $254 - z^3 = 46$ |
|---|---|---|
|  |  |  |

5. *(Challenge.)* These equations involve cube roots of negative numbers. Give each solution in exact form.

| a. $x^3 = -27$ | b. $w^3 = -343$ | c. $4t^3 = -4$ |
|---|---|---|
|  |  |  |
| d. $5x^3 + 3 = -27$ | e. $-10r^3 = 10\,000$ | f. $54 - x^3 = -1$ |
|  |  |  |

6. Here are some more practice problems. Round the answers to three decimals.

| a. $45 - x^2 = 20$ | b. $112^2 + s^2 = 18\,200$ | c. $6650 - y^2 = 70^2$ |
|---|---|---|
|  |  |  |

**Puzzle Corner**   Solve $x^2 - x = 0$.

# The Pythagorean Theorem

You will now learn a very famous mathematical result, the Pythagorean Theorem, which has to do with the lengths of the sides in a right triangle. First, we need to study some terminology.

In a right triangle, the two sides that are perpendicular to each other are called **legs**. The third side, which is always the longest, is called the **hypotenuse**.

In the image on the right, the sides $a$ and $b$ are the legs, and $c$ is the hypotenuse.

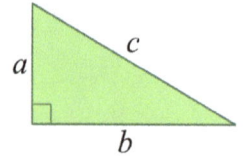

Note: We don't use the terms "leg" and "hypotenuse" to refer to the sides of an acute or obtuse triangle — this terminology is restricted to *right* triangles.

The Pythagorean Theorem states that **the sum of the squares of the legs equals the square of the hypotenuse.**

In symbols it looks much simpler:

$$a^2 + b^2 = c^2$$

The picture shows squares drawn on the legs and on the hypotenuse of a right triangle. Verify visually that the total area of the two yellow squares drawn on the legs looks about equal to the area of the blue square on the hypotenuse.

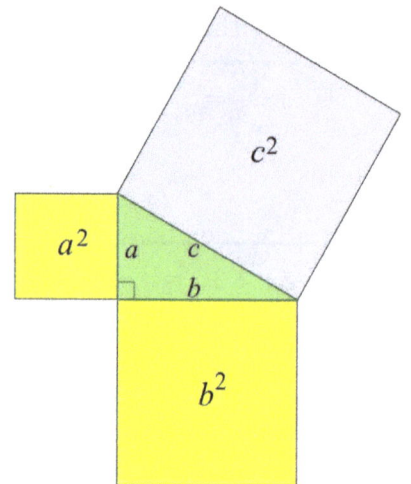

We will prove this theorem in another lesson.
For now, let's get familiar with it and learn how to use it.

1. Below you see the famous 3-4-5 triangle: its sides measure 3, 4, and 5 units and it is a right triangle. Check that the Pythagorean Theorem holds for it by filling in the numbers on the right.

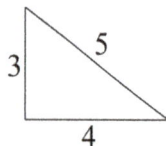

2. If we double the lengths from the 3-4-5 triangle, we get the 6-8-10 triangle. Use a ruler and a protractor to draw line segments that are 6 cm and 8 cm long and are perpendicular to each other.

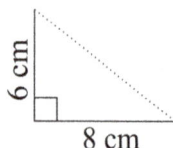

   Then draw in the third side to complete the triangle. Does the hypotenuse of your triangle measure 10 cm? Do the lengths 6, 8, and 10 fulfill the Pythagorean Theorem?

88

**Example 3.** The two legs of a right triangle measure 7 and 10 units. How long is the hypotenuse?

Let $x$ be the length of the unknown side, which is the hypotenuse. From the Pythagorean Theorem, we get:

$$7^2 + 10^2 = x^2$$
$$49 + 100 = x^2$$
$$x^2 = 149$$
$$x = \sqrt{149} \quad \text{or } x = -\sqrt{149}$$

We ignore the negative root as the length of a side cannot be negative!

The answer is left in the root form since this is not a real-life problem.

3. Solve for the unknown side of each right triangle. Leave your answer in root form if the radicand (number under the radical) is not a perfect square.

a.

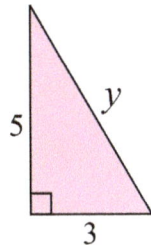

b.

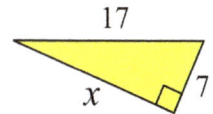

c.

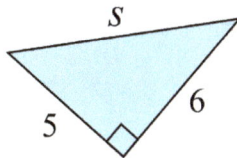

d.

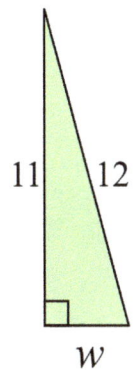

4. The sides of a square measure 6 units. How long is the diagonal of the square?
   Give your answer to two decimal digits.

**Example 4.** The hypotenuse of a right triangle is $\sqrt{27}$ and its one leg is 3. How long is the other leg?

Let $y$ be the length of the unknown leg. According to the Pythagorean Theorem, we get:

$$y^2 + 3^2 = (\sqrt{27})^2$$
$$y^2 + 9 = 27$$
$$y^2 = 18$$
$$y = \sqrt{18}$$

Again, we do not include the negative root since the length of a side cannot be negative.

Being a purely mathematical problem, the answer is left in the root form.

5. Solve for the unknown side of each right triangle. Leave your answer in root form if the radicand (number under the radical) is not a perfect square.

**a.**

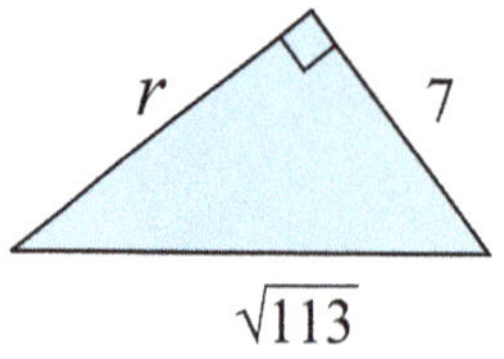

$r$    7    $\sqrt{113}$

**b.**

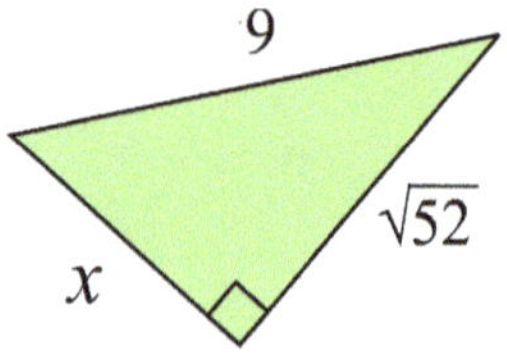

9    $\sqrt{52}$    $x$

**c.**

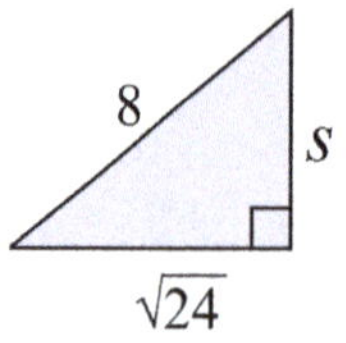

8    $s$    $\sqrt{24}$

**d.**

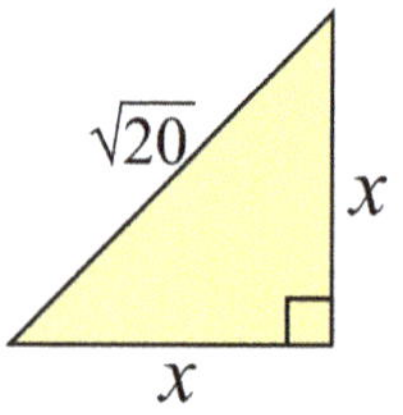

$\sqrt{20}$    $x$    $x$

6. **a.** The two legs of a right triangle are $\sqrt{7}$ and $\sqrt{8}$.
How long is the hypotenuse?

**b.** The hypotenuse of a right triangle is $\sqrt{67}$ and one leg is $\sqrt{41}$.
How long is the other leg?

**Example 5.** Find the unknown leg of this right triangle.

This time we know the hypotenuse and one of the legs. The Pythagorean Theorem gives us:

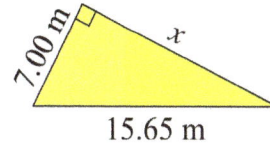

$$7.00^2 + x^2 = 15.65^2$$

$$49 + x^2 = 244.9225$$

$$x^2 = 195.9225$$

$$x = \sqrt{195.9225} \text{ or } x = -\sqrt{195.9225}$$

$$x \approx 14.00 \text{ m}$$

We keep all the decimals for the intermediate results.

Again, we ignore the negative root.

The answer is given to the hundredth of a metre, just like the lengths given in the problem.

The other leg measures about 14.00 m.

7. Solve for the unknown side. Round your answer to the same accuracy as the numbers in the problem.

a.

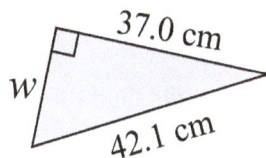

b.

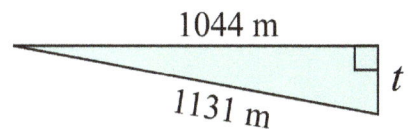

c.

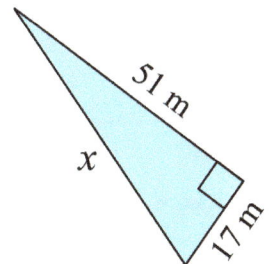

8. If the legs of a right triangle measure 149 cm and 92 cm, find the length of the hypotenuse to the nearest centimetre.

9. **a.** Measure each side of this triangle to the nearest millimetre.

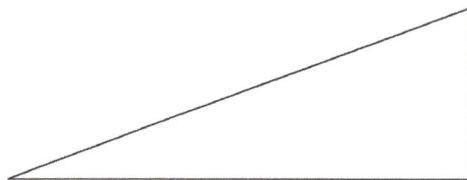

   **b.** Verify that the sum of the areas of the squares on the legs is *very close* to the area of the square on the hypotenuse. I say "very close" because the process of measuring is always inexact, and therefore your calculations and results will probably not yield true equality, just something close.

10. How long is the diagonal of a laptop screen that is 9.0 inches high and 14.4 inches wide?

   Note: when a laptop is advertised as having a "15-inch screen," it is the *diagonal* that is 15 inches, not the width or the height.

---

**Puzzle Corner**    A math teacher made the problem below for a test. Find what went wrong with it. Then fix the problem, so it can be used in the test, and solve it.

How long is the unknown side?

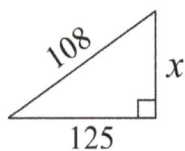

---

# Applications of the Pythagorean Theorem 1

**Example 1.** A four-metre ladder is placed against a wall so that the base of the ladder is 1 metre away from the wall. What is the height of the top of the ladder?

Since the ladder, the wall, and the ground form a right triangle, this problem is easily solved by using the Pythagorean Theorem. Let $h$ be the unknown height. From the Pythagorean Theorem, we get:

$$1^2 + h^2 = 4^2$$
$$1 + h^2 = 16$$
$$h^2 = 15$$
$$h = \sqrt{15}$$
$$h \approx 3.87$$

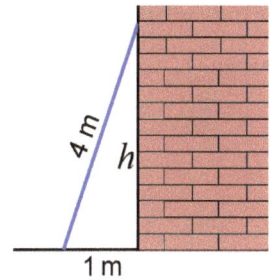

Our answer, 3.87, is in metres. This means the ladder reaches to about 3.87 metres high.

1. The area of a square is 100.0 m². How long is the diagonal of the square?

2. A park is in the shape of a rectangle and measures 48 m by 30 m. How much longer is it to walk from A to B around the park than to walk through the park along the diagonal path?

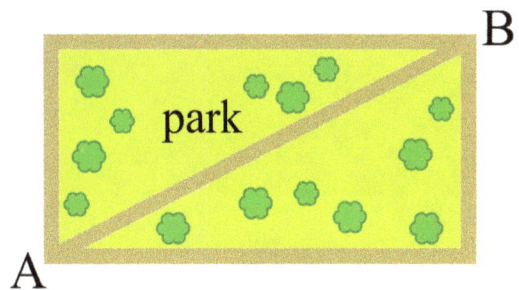

3. A clothesline is suspended between two apartment buildings.
   Calculate its length, assuming it is straight and doesn't sag any.

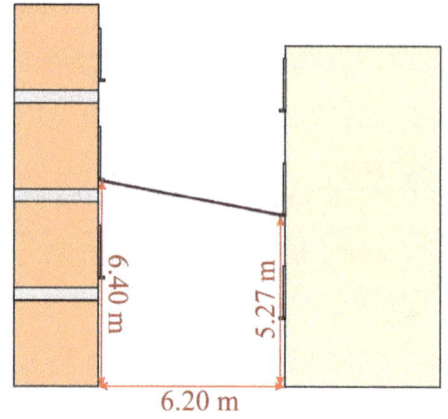

4. Find the perimeter of triangle ABC to the nearest
   tenth of a unit.

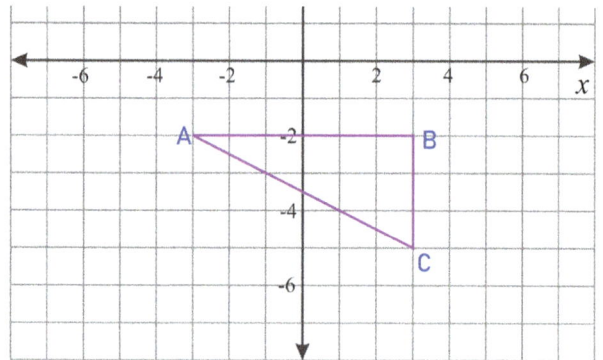

**Example 2.** Find the area of an isosceles triangle with sides 92 cm, 92 cm, and 40 cm.

**Solution:** To calculate the area of any triangle, we need to know its altitude. When we draw the altitude, we get a right triangle:

The next step is to apply the Pythagorean Theorem to solve for the altitude $h$, and after that calculate the actual area.

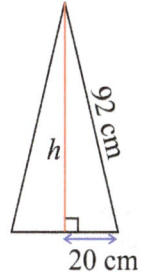

5. Calculate the area of the isosceles triangle in the example above to the nearest ten square centimetres.

6. Calculate the area of an equilateral triangle with 24-cm sides to the nearest square centimetre. Don't forget to draw a sketch.

# A Proof of the Pythagorean Theorem and of Its Converse

There exist hundreds of different proofs for the Pythagorean Theorem. In this lesson, we will look at two geometric proofs.

**Proof.**

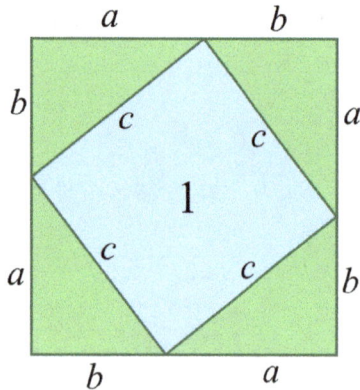

The figure above has four right triangles, each with sides $a$, $b$ and $c$. The sides of the outside square are $a + b$. The triangles enclose a square with sides $c$ units long.

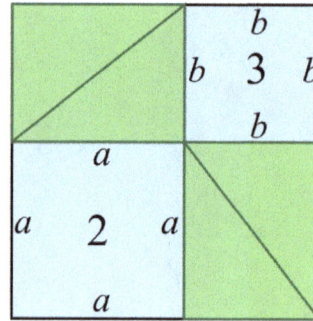

Here the sides of the large square are still $a + b$, but the four right triangles have been rearranged so that two smaller squares are formed, one with side $a$ and the other with side $b$.

Since the areas of both large squares are equal, and the areas of the four right triangles are equal, it follows that the remaining (blue) areas are also equal. In other words, the area of square 1, which is $c^2$, equals the area of square 2 (which is $a^2$) plus the area of square 3 (which is $b^2$). In symbols, $c^2 = a^2 + b^2$. ☺

1. Figure out how this proof of the Pythagorean Theorem works.

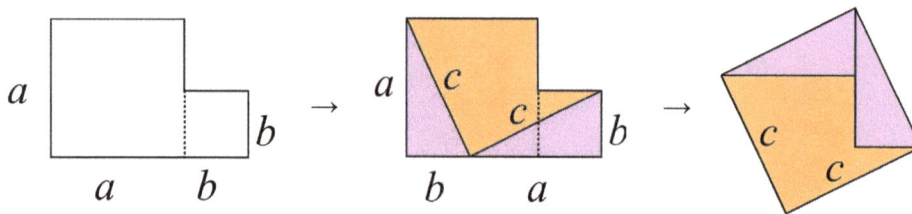

2. Study one of the proofs enough so you can explain it to someone else. Then do so.

Often in mathematics, **theorems** are of the form "If (statement 1), then (statement 2)."
The **converse** of such a theorem is, "If (statement 2), then (statement 1)."

For example, here is a true theorem that follows this "*if-then*" format:

"If the corresponding angles in two triangles are congruent, then the triangles are similar."

Its converse is: "If two triangles are similar, then their corresponding angles are congruent" , which is also true.

But sometimes, a mathematical statement is only true in one way, and its converse is not true. Here is an example of that.

Theorem: "If the sides of two rectangles are congruent, their areas are equal." (true)

Its converse: "If two rectangles have equal areas, their sides are congruent." (not true)

If we write the Pythagorean Theorem in this format of "If (statement 1), then (statement 2)", it becomes:

If a triangle with sides $a$, $b$, and $c$ is a right triangle, then its sides fulfil the equation $a^2 + b^2 = c^2$.

Its converse is:

If the sides $a$, $b$, and $c$ of a triangle fulfil the equation $a^2 + b^2 = c^2$, then the triangle is a right triangle.

3. Tell the converse of each statement. Then tell whether the original statement and its converse are true or not.

**a.** STATEMENT: If there are no clouds in the sky, then it is not raining.　　　　true / false

　　CONVERSE:　　　　true / false

**b.** STATEMENT: If you eat spoiled food, you will get gastrointestinal symptoms.　　　　true / false

　　CONVERSE:　　　　true / false

**c.** STATEMENT: If you just had your 7th birthday, then you'll turn 8 in a year.　　　　true / false

　　CONVERSE:　　　　true / false

**d.** STATEMENT: If $a + b = c$, then $c - b = a$.　　　　true / false

　　CONVERSE:　　　　true / false

**e.** STATEMENT: If you multiply two odd numbers, the product is also odd.　　　　true / false

　　CONVERSE:　　　　true / false

The **converse of the Pythagorean Theorem** states that if the relationship $c^2 = a^2 + b^2$ is true for the side lengths $a$, $b$, $c$ of some triangle, then the triangle is a right triangle.

**Proof.** This is a proof by contradiction. This means that we start by assuming the statement we're trying to prove is *not* true, and then try to derive a contradiction. If we succeed, then it follows that the original statement must be true.

Let's say that the relationship $c^2 = a^2 + b^2$ is true for the side lengths $a$, $b$, $c$ of triangle ABC, but the triangle is *not* a right triangle.

Let's further assume the angle at A is an acute angle. (The proof will work almost identically if it was obtuse.)

Now, we will draw a line segment of the length $a$ so that the angle BAC′ is a right angle, as in the image below:

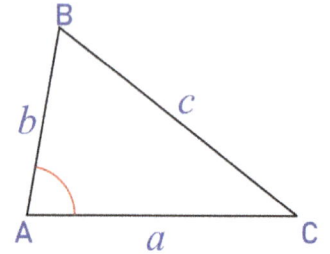

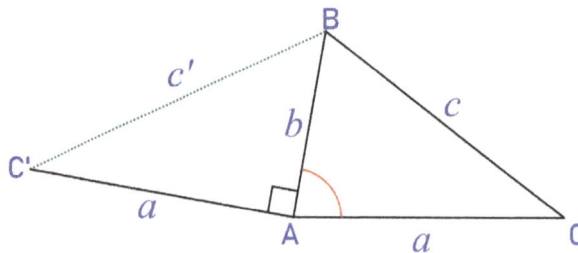

This forms the *right* triangle ABC′, which shares the side $\overline{AB}$ with triangle ABC. The side lengths of this triangle are $a$, $b$, and $c'$.

Now, since triangle ABC′ is a right triangle, the Pythagorean Theorem holds, and thus, $a^2 + b^2 = c'^2$. And our original assumption was that $a^2 + b^2 = c^2$.

From those two equations, it follows that $c^2 = c'^2$, and since $c$ and $c'$ cannot be negative, it follows that $c = c'$. So, the two triangles have the same lengths of sides: $a$, $b$, and $c$.

Yet, with three given lengths, you can only make *one* unique triangle, not two that would have different angles. This is a contradiction. Thus, our assumption that ABC is not a right triangle cannot be true.

Hence, ABC is a right triangle, and the converse of the Pythagorean Theorem is true.

4. Study the proof and make sure you understand it. Study it enough that you can tell it to someone else. Then tell it to someone else.

5. Is this corner a right angle?

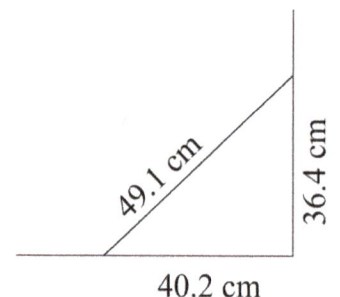

49.1 cm

36.4 cm

40.2 cm

6. Construction workers have made a (hopefully) rectangular mold out of wood, and they are getting ready to pour cement into it. How could they make sure that the mold is indeed a rectangle and not a parallelogram? After all, in a parallelogram the opposite sides are equal, so simply measuring the opposite sides does not guarantee that a shape is a rectangle.

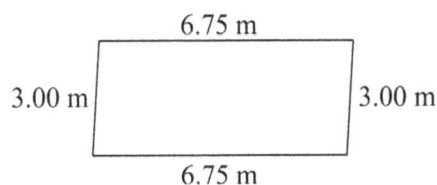

6.75 m

3.00 m     3.00 m

6.75 m

---

**Example 1.** (optional) This triangle is *not* a right triangle, so the Pythagorean Theorem does *not* hold:

$$2.55^2 + 3.31^2 \overset{?}{=} 3.58^2$$

$$6.5025 + 10.9561 \overset{?}{=} 12.8164$$

$$17.4586 > 12.8164$$

(In the image, "$u^2$" signifies a square unit.)

The sum of the areas of the squares drawn on the two shortest sides is *more* than the area of the square drawn on the longest side. As you can see, the triangle is acute.

If $a^2 + b^2 < c^2$ (where c is the longest side), then the triangle is obtuse.

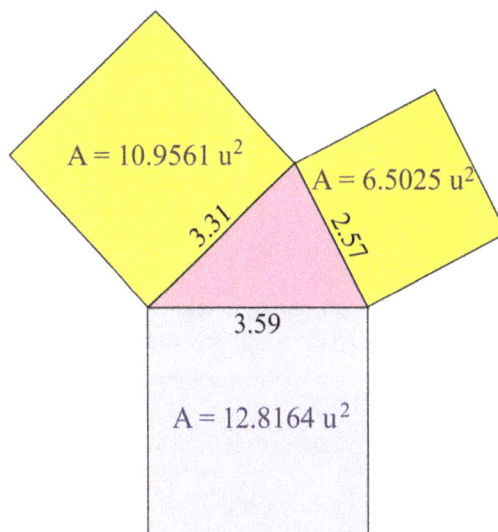

$A = 10.9561\ u^2$

$A = 6.5025\ u^2$

3.31     2.57

3.59

$A = 12.8164\ u^2$

---

7. (optional) For each set of lengths below, determine whether the lengths form an acute, right, or obtuse triangle. You can use the Pythagorean theorem, and/or construct the triangles using a compass and a ruler. To learn or to review how to do the latter, check this video — the fifth example in it explains how to draw a triangle with three given sides: https://www.mathmammoth.com/videos/geometry/draw_triangles

| **a.** 6, 9, 13 | **b.** 12, 13, 5 | **c.** 4, 5, 7 |
|---|---|---|
| **d.** 4, 5, 6 | **e.** 13, 11, 10 | **f.** 15, 20, 25 |

# Applications of the Pythagorean Theorem 2

**Example 1.** Find the diagonal of the rectangular prism on the right.

The diagonal in question is marked with $x$. To find its length, we will use the right triangle shaded in pink.

That right triangle, in its turn, has as one of its legs the diagonal ($d$) of the bottom square of the prism.

We will first solve for $d$ using the Pythagorean Theorem:

$$3^2 + 3^2 = d^2$$
$$18 = d^2$$
$$d = \sqrt{18}$$

We don't need a decimal approximation for $d$, since this is only an intermediary result. (Since we continue the calculation, it is better to use the exact value, but if not, you would want to keep at least 4 decimals.)

Next, we look at the right triangle with sides 5, $d$ or $\sqrt{18}$, and $x$, and use the Pythagorean Theorem. This is left for you to do in exercise 1.

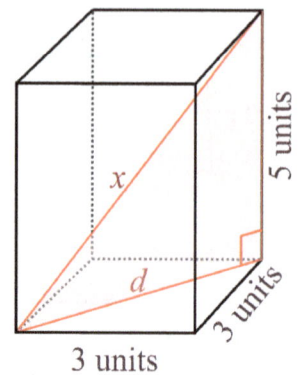

1. Finish solving the problem in Example 1. Give your answer in root form.

2. Find the diagonal of a cube with 15-cm edges.
   Draw a sketch first.

3. The picture shows a right pyramid with a 12-cm square as the base. Find its surface area. (You will need to draw additional lines to the picture.)

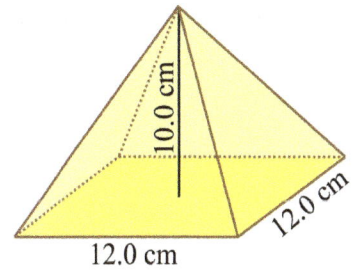

4. Calculate the length of the rafter in feet, if...

   **a.** ...the run is 12 ft and the rise is 3 ft

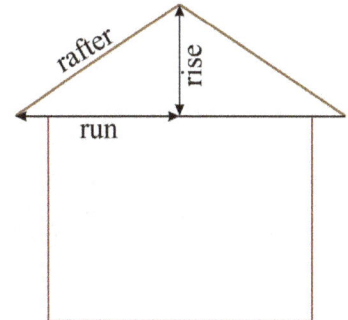

   **b.** ...the run is 12 ft and the rise is 5.25 ft.

5. Find the surface area of this roof to the nearest tenth of a square metre.

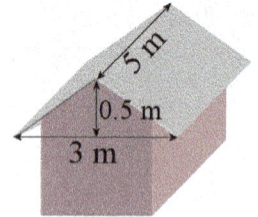

6. The roof of a little kiosk is in the shape of a square pyramid. Each bottom edge measures 3.5 m, and the other edges measure 3.2 m. Find the surface area of the roof to the nearest tenth of a square metre.

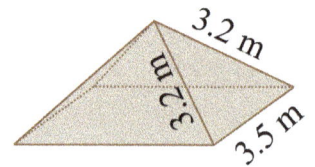

7. A creek runs through a piece of land in a straight line.

   a. Find the length of the creek. Give your answer to
      the same accuracy as the dimensions in the picture.

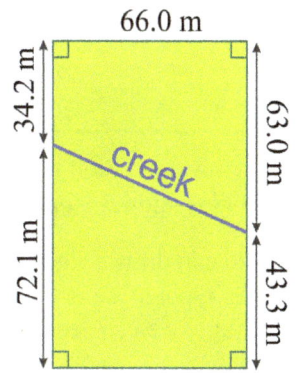

   b. The creek splits the plot into two parts. Calculate the areas
      of the two parts.

The hypotenuse of a right triangle measures 12.0 m, and the one leg is
twice the length of the other. Find the side lengths of the triangle.

# Distance Between Points

**Example 1.** Find the distance between the points $(-3, 5)$ and $(2, -1)$.

We can draw a right triangle so that the length of the hypotenuse is the desired distance. From the image, we can see the legs are 5 and 6 units long.

Now it is easy to use the Pythagorean Theorem to find the distance:

$$5^2 + 6^2 = x^2$$
$$25 + 36 = x^2$$
$$61 = x^2$$
$$x = \sqrt{61} \approx 7.8 \text{ units}$$

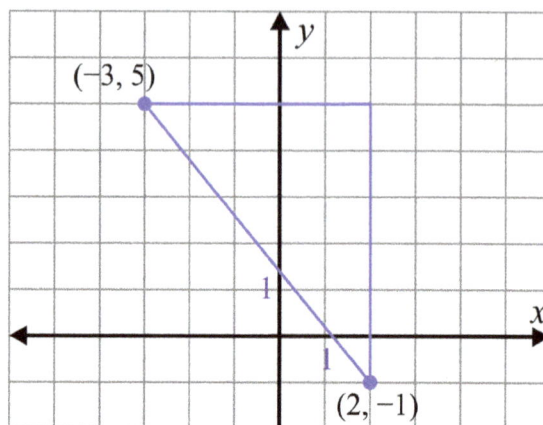

1. Find the distance between the points. Give your answer both in an exact form, and rounded to one decimal digit.

   **a.** $(4, 7)$ and $(-5, 2)$

   **b.** $(4, -3)$ and $(-7, 0)$

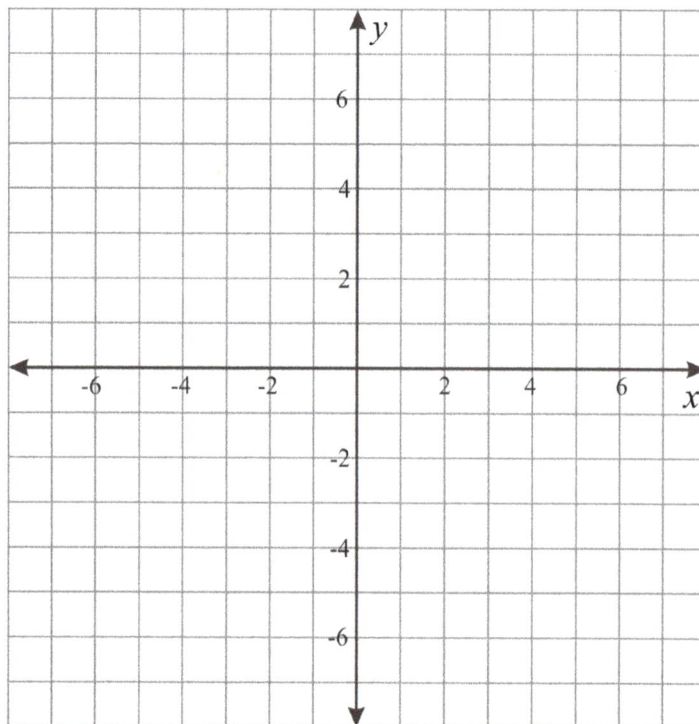

2. Find the distance between $(-2, 0)$ and $(8, 11)$ without drawing the points on the grid. Give your answer both in an exact form, and rounded to one decimal digit.

3. Explain how to find the distance between (5, 2) and (30, 45), and also find the distance.
   *Hint*: consider the horizontal distance and the vertical distance between the points.
   Give your answer in an exact form.

4. Find the distances between the points using the Pythagorean Theorem. Give your answer in an exact form.

| a. (−10, 9) and (22, 15) | b. (30, −25) and (−7, −32) |
|---|---|
| | |

5. Find the perimeter of the triangle with vertices at (−4, −4), (−4, 5), and (2, 5), to the nearest tenth of a unit.

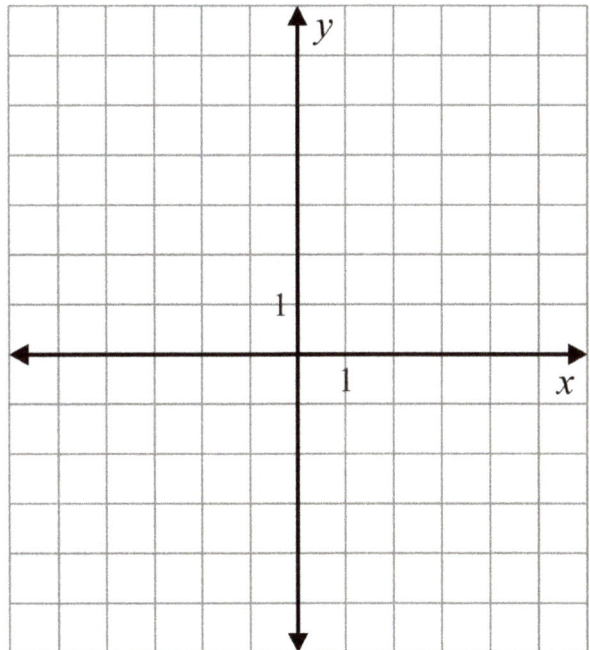

6. Find the area of the square with vertices at (−5, 0), (0, 5), (5, 0), and (0, −5) to the nearest tenth of a square unit.

7. Which shape has a shorter perimeter? How much shorter (to the nearest tenth of a unit)?

8. How much bigger is the area of trapezoid DEFG than the area of triangle ABC?

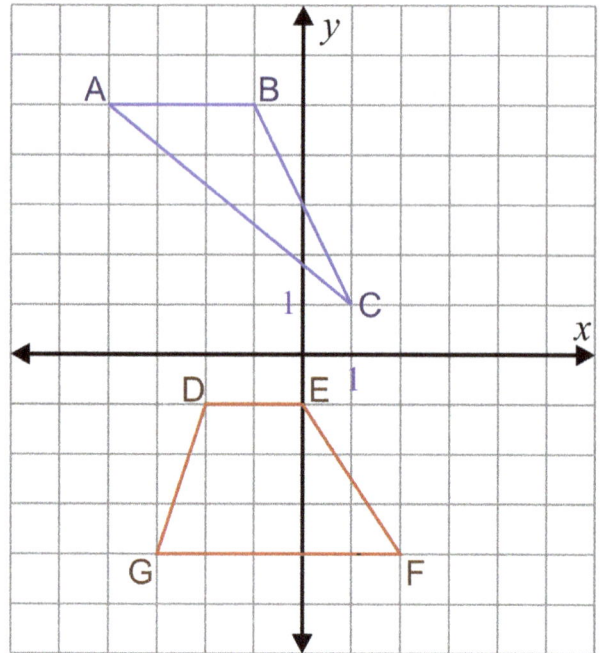

9. On a map, Sarah's home is at the origin, and the local swimming pool is at (50, 32). To go there, Sarah has to follow the roads, and travel first to the point (50, 0), then to (50, 32), whereas a crow can fly from Sarah's home directly to the pool. If each unit on the map is 10 meters, how much shorter is the way a crow flies than the way Sarah has to travel?

Puzzle Corner   Find the area of a regular octagon with sides 2 units long.

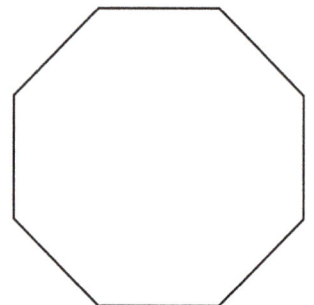

106

# Mixed Review Chapter 6

1. Write an equivalent expression using the exponent laws, without negative exponents.

| | | | |
|---|---|---|---|
| **a.** $3x^4 y^5 y^2 \cdot 6x^6 =$ | **b.** $(3x)^{-3} =$ | **c.** $(3yz)^2$ | **d.** $(b^{-2})^4 =$ |
| **e.** $\dfrac{8x^5}{28x^8} =$ | **f.** $\dfrac{x^{-5}}{x^2} =$ | **g.** $\left(\dfrac{-2}{5y}\right)^2 =$ | **h.** $\left(\dfrac{3s}{t^2}\right)^4 =$ |

2. Draw a dilation of triangle ABC...

| **a.** from point A with scale factor 1/3 | **b.** from point C with scale factor 1/2 |
|---|---|
|  |  |

3. Eight tennis balls fit snugly in a cube-shaped container. Calculate what fraction of the total volume of the cube the tennis balls take up.
*Hint: write this fraction using the formulas for the volumes, and simplify it.*

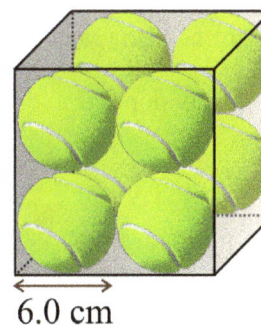

6.0 cm

4. Chloe bought 10 metres of material, five metres at the regular price of $5.95/m and the rest at some discounted price. Her total came to $53. At home, she started wondering how much the discount was.

Write an equation to solve what the unknown discounted price was. Use $p$ for the discounted price. Then solve your equation.

5. **a.** Make up two functions for the cost of renting a surfboard as a function of time. The first should be a proportional relationship, and the other nonlinear. Make your functions reasonable so that the cost of renting a surfboard for an entire day (8 hours) is $50 at a maximum.

Give the linear function as an equation, and the nonlinear one as a table of values.

Function 1:

Function 2:

| time (hours) | 0 | 1 | 2 | 3 | 4 | 5 | 6 | 7 | 8 |
|---|---|---|---|---|---|---|---|---|---|
| Cost ($) | | | | | | | | | |

**b.** Which function gives a better deal if you are renting a surfboard for 2 hours?   For 6 hours?

6. Find the equation of each line, in slope-intercept form. Also graph the lines.

**a.** has slope −2 and passes through (−2, 6)

**b.** has slope 2/3 and passes through (4, −4)

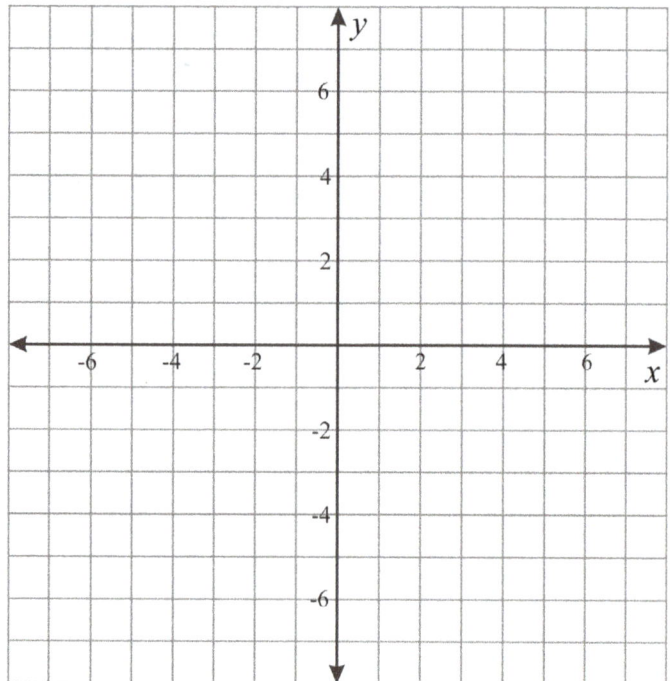

7. **a.** What is the equation of a horizontal line that passes through $(-3, -5)$?

**b.** What is the equation of a vertical line that passes through $(9, 8)$?

**c.** What is the equation of a line that is parallel to $y = 5x + 2$ and passes through $(1, 2)$?

8. Convert, rounding your answer to the same number of significant digits as the measurement.

**a.** 71.0 cm = _____ in     **b.** 2 400 kg = _____ lb

                                                                     1 inch = 2.54 cm

**c.** 235 ft = _____ m     **d.** 83.5 lb = _____ kg

                                                                      1 ft = 0.3048 m

**e.** 15.69 m = _____ ft     **f.** 4.5 in = _____ cm

                                                                     1 kg = 2.2 lb

9. If 3 cm of rain falls over one square kilometre, how many raindrops fell? Give your answer in scientific notation, to three significant digits.

Besides the well-known conversion factors, here are some facts you may need:

- One cubic metre = 1000 litres.
- The size of raindrops varies but for this problem, use $2.65 \cdot 10^4$ raindrops per litre.

# Chapter 6 Review

1. Find the values of these (principal) square roots and cube roots.

| a. $\sqrt{64}$ | b. $\sqrt{169}$ | c. $\sqrt{2500}$ | d. $\sqrt{0.81}$ |
|---|---|---|---|
| e. $\sqrt{\dfrac{36}{100}}$ | f. $-\sqrt{49}$ | g. $\sqrt[3]{125}$ | h. $\sqrt[3]{27\,000}$ |

2. Between which two whole numbers do the following square roots lie? Do not use a calculator.

a. $\sqrt{7}$        b. $\sqrt{77}$        c. $\sqrt{134}$        d. $\sqrt{48}$

3. Find the value of $\sqrt{13}$ to one decimal digit, without a calculator.

4. Plot the following numbers *approximately* on the number line. Do not use a calculator, but think between which two integers the square root lies, and whether it is close to one of those integers, using mental math.

$\sqrt{50}/2$      $\sqrt{17}$      $-\sqrt{8}$      $\sqrt[3]{8}$      $\sqrt{101}-4$      $-\sqrt[3]{27}$      $\pi$    $-\sqrt{38}/3$

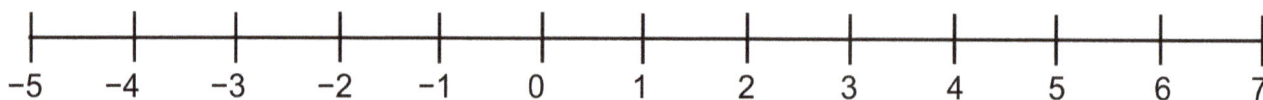

```
 ├──┼──┼──┼──┼──┼──┼──┼──┼──┼──┼──┼──┼──┤
-5   -4   -3   -2   -1    0    1    2    3    4    5    6    7
```

5. Compare, writing >, <, or = between the numbers. Think between which two whole numbers the square root lies, using mental math.

| a. $11$ ☐ $\sqrt{150}$ | b. $\sqrt{76}$ ☐ $9$ | c. $\sqrt{20}$ ☐ $4$ | d. $\sqrt[3]{10}$ ☐ $2$ |
|---|---|---|---|
| e. $4$ ☐ $\pi+1$ | f. $\sqrt{85}/3$ ☐ $3$ | g. $\sqrt{27}+2$ ☐ $6$ | h. $\sqrt{68}-3$ ☐ $6$ |

6. Find the value of the expressions.

| a. $\sqrt{144}$ | b. $-\sqrt{81}$ | c. $\sqrt{1600}$ |
|---|---|---|
| d. $\sqrt{10^2 - 6^2}$ | e. $\sqrt{49 \cdot 49}$ | f. $\sqrt{5 \cdot (83 - 3)}$ |

7. **a.** If the side of a square measures $\sqrt{7}$, what is its area?

    **b.** What is the perimeter of a square with an area of 20 square units?
        Give your answer as an exact value (not rounded).

8. Determine whether the following numbers are rational or irrational, and explain why.

    **a.** 0.8053                           **b.** $\sqrt{2500}$

    **c.** $5\pi$                              **d.** $-\sqrt{56}$

    **e.** $-\dfrac{2}{7}$                        **f.** $\dfrac{1}{\sqrt{36}}$

    **g.** $2.1\overline{09}$                      **h.** 0.020202

    **i.** $0.20\overline{8}$                       **j.** $4\sqrt{7}$

9. Write each repeating decimal as a fraction.

| a. $0.\overline{61}$ | b. $4.1\overline{7}$ |
|---|---|
| | |

10. Solve. Give the final answers in exact form.

| a. $\quad x^2 \;=\; 147$ | b. $\quad a^2 \;=\; 169$ | c. $\quad w^3 \;=\; 0.36$ |
|---|---|---|
| d. $\quad 3x^3 \;=\; 21$ | e. $\quad 5b^3 \;=\; 625$ | f. $\quad 2a^3 \;=\; -16$ |

11. Solve. Give your answer to the nearest thousandth. You may use a calculator.

| a. $\quad y^2 + 18 \;=\; 35$ | b. $\quad 0.6h^2 \;=\; 4$ |
|---|---|
| | |

12. For each set of lengths, determine whether they form a right triangle.

    **a.** 20, 24, 30

    **b.** 2.6, 1.0, 2.4

13. Solve for the unknown side. Leave your answer in root form if the radicand is not a perfect square.

| **a.** | **b.** |
|---|---|
| | |

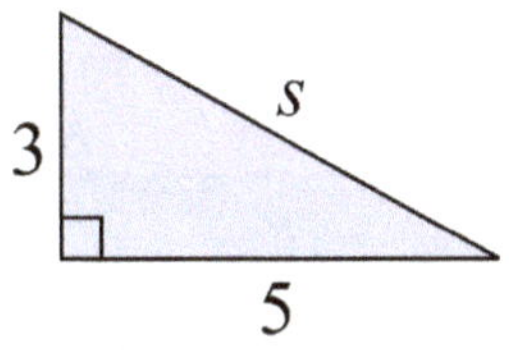

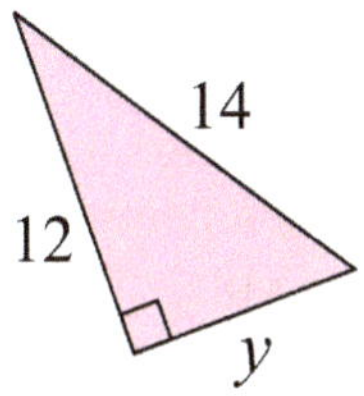

14. Solve for the unknown side. Round your answer to the same accuracy as the given numbers.

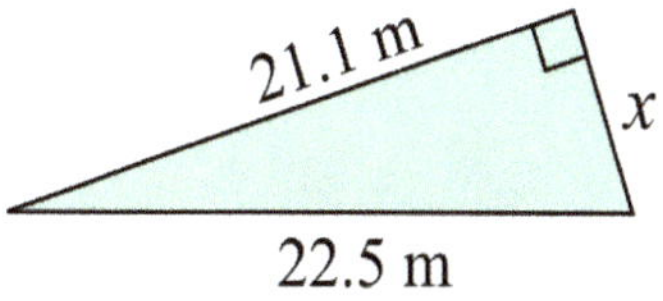

15. The two legs of a right triangle are $\sqrt{7}$ and $\sqrt{8}$.
    How long is the hypotenuse?

16. Lauren and Anna want to make this pennant for their jogging club. Calculate its area.

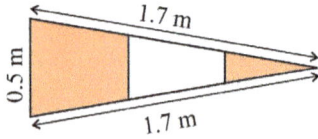

17. Arrange the pieces in the empty square
    in a manner that will prove the
    Pythagorean Theorem, and explain
    how your arrangement does so.

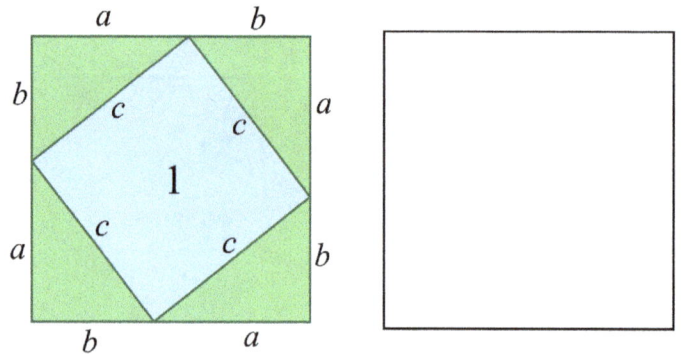

18. Find the area and the perimeter of the garden,
    if one unit in the grid is 0.50 metres.

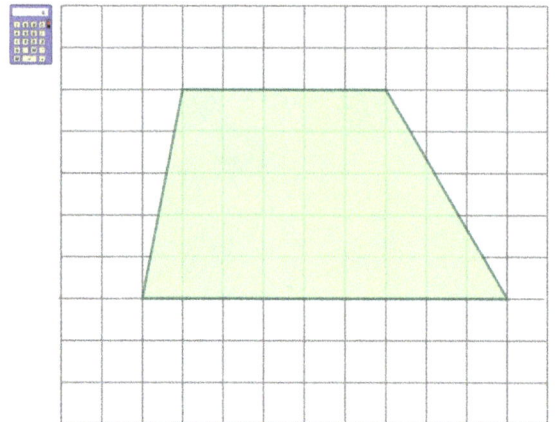

19. Find the distances between the points, to the nearest tenth of a unit.

| **a.** $(-3, 4)$ and $(11, -9)$ | **b.** $(42, -15)$ and $(70, 100)$ |
|---|---|
| | |

20. Find the diagonal of the cube with edges 2.5 m long.

21. The map shows part of downtown Nashville, Tennessee. The triangle ABC on the map is very close to a right triangle. The distance AB is 370 m and the distance AC is 620 m. However, these distances are approximate, so your calculations will also be only approximate.

About how much shorter is it to travel from point A to point C along Lafayette Street than to travel first along Korean Veterans Boulevard and then along 5th Avenue South?

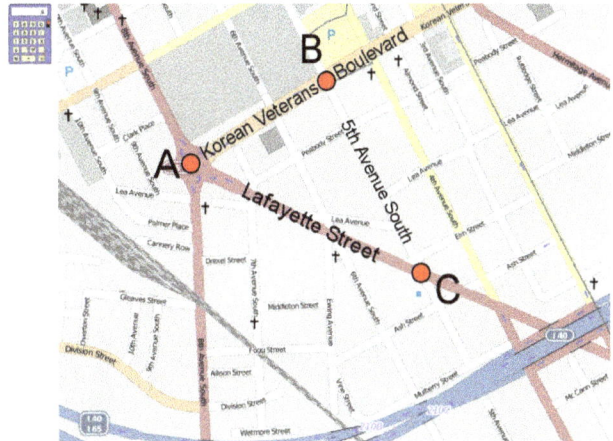

# Chapter 7: Systems of Linear Equations
## Introduction

This chapter covers how to analyze and solve pairs of simultaneous linear equations. (The equations studied contain only two variables.)

The first lesson, *Equations with Two Variables*, is optional. It reinforces the idea that a point on a line satisfies the equation of the line, and thus prepares the way for the main topics in the book.

First, students learn to solve systems of linear equations by graphing. Since each equation is an equation of a line, this is a simple technique, but it has its limitations, thus, algebraic solution methods will be taught also, in later lessons.

In the next lesson, we look at the number of solutions that a system of two linear equations can have. The three possible situations are easy to see based on the graphs of the equations: either one solution (the lines intersect in one point), no solutions (lines are parallel), or an infinite number of solutions (the lines are the same).

In the next lesson, students learn the algebraic method of solving systems of equations by substitution. This is a straightforward technique that many students will grasp easily. However, one has to be careful not to make simple mistakes. The lesson has some practice problems where students practise finding errors in solutions. Instruct the student(s) to check their solutions each time, as that is the best way to catch errors.

As for me (Maria, the author), as I wrote the answer key, I immediately checked my solution for each system of equations, and several times found an error that way. (The funniest errors were when I had switched from $x$ to $y$ in the middle of the solution!) So, checking the solution is important. To save space the answer key does not include the checks, but the student should always do that, whether with mental math or with a calculator.

The following lesson, *Applications, Part 1*, has a variety of word problems that students can now solve using a system of equations.

After that, students learn another algebraic method for solving systems of equations: the addition or elimination method. This is useful when the coefficients of the variables are such that you can easily find their least common multiple. Students also practise solving more complex systems, where the equations first have to be transformed and simplified, or include fractions.

Then it is time for more word problems, in the lesson *Applications, Part 2*. One lesson is devoted to problems about speed, time, and distance, and another for mixtures and comparisons. Making a chart is very helpful in these situations.

## Pacing Suggestion for Chapter 7

This table does not include the chapter test as it is found in a different book (or file).
Please add one day to the pacing if you use the test.

| The Lessons in Chapter 7 | page | span | suggested pacing | your pacing |
|---|---|---|---|---|
| Equations with Two Variables | 119 | *4 pages* | 1 day | |
| Solving Systems of Equations by Graphing | 123 | *5 pages* | 1-2 days | |
| Number of Solutions | 128 | *4 pages* | 1 day | |
| Solving Systems of Equations by Substitution | 132 | *7 pages* | 2 days | |
| Applications, Part 1 | 139 | *4 pages* | 1 day | |
| The Addition Method, Part 1 | 143 | *5 pages* | 1 day | |
| The Addition Method, Part 2 | 148 | *5 pages* | 1 day | |
| More Practice | 153 | *4 pages* | 1 day | |

## Helpful Resources on the Internet

We have compiled a list of Internet resources that match the topics in this chapter, including pages that offer:

- **online practice** for concepts;
- online **games**, or occasionally, printable games;
- **animations** and interactive **illustrations** of math concepts;
- **articles** that teach a math concept.

We heartily recommend you take a look! Many of our customers love using these resources to supplement the bookwork. You can use these resources as you see fit for extra practice, to illustrate a concept better and even just for some fun. Enjoy!

https://l.mathmammoth.com/gr8ch7

Scan me

# Equations with Two Variables

## (This lesson is optional.)

The equation $2x + 3y = 16$ has **two variables**, $x$ and $y$. One solution to the equation is $x = 2$ and $y = 4$, because when we substitute those values to the equation, it checks, or is a true equation:

$$2(2) + 3(4) = 16$$

But it also has the solution $x = 0.5$ and $y = 5$:

$$2(0.5) + 3(5) = 16$$

In fact, we can choose any number we like for the value of $x$, and then *calculate* the value of $y$, and thus find another solution to the equation.

For example, if we choose $x = -1$, then we get

$$2(-1) + 3y = 16$$

from which $y = (16 + 2)/3 = 6$. So, $x = -1$, $y = 6$ is yet another solution.

All of these solutions, having both $x$ and $y$ values, are **number pairs**, and can be considered as **points on the coordinate plane.**

We can make a table of some of the possible $(x, y)$ values (solutions):

| $x$ | $y$ |
|-----|-----|
| $-1$ | 6 |
| 0 | 16/3 |
| 0.5 | 5 |
| 2 | 4 |

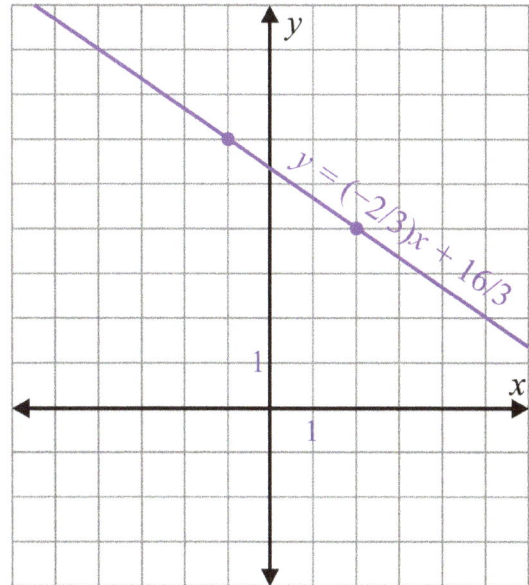

...and there are many more. When plotted, **these points fall on a line** — and you can probably guess, the equation of that line is $2x + 3y = 16$!
(Or, in slope-intercept form, $y = (-2/3)x + 16/3$.)

A line in the coordinate plane represents all the solutions to the equation that is the equation of the line. In other words, **each point on the line is a solution to the equation.**

1. Find three solutions to the equation $5x + 2y = 32$.

2. Find three solutions to the equation $-4x + y = -6$.

3. **a.** What is the equation if its solution set is
represented by this line?

**b.** List two distinct integer number pairs that
are solutions to the equation.

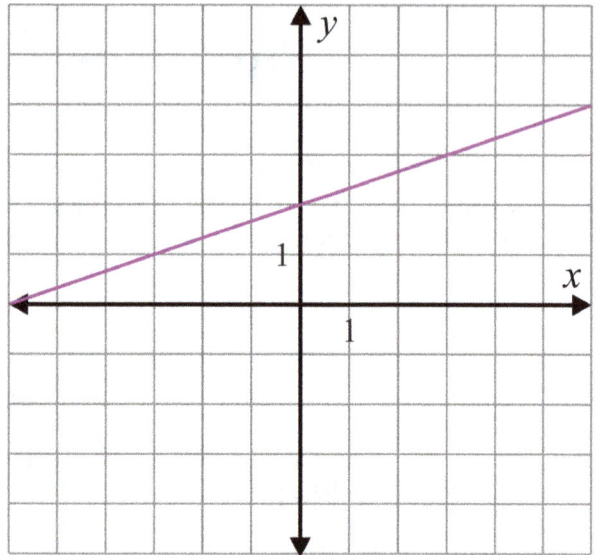

4. A certain linear equation with two variables has as solutions $(0, -5)$, $(2, 3)$ and $(4, 11)$. Find the equation.

5. A certain linear equation with two variables has as solutions $(-1, -5)$ and $(2, 8)$. Find the equation.

6. Party hats cost $2 apiece and party whistles cost $3 apiece. Randy bought $x$ hats and $y$ whistles.

   **a.** Write an expression depicting the total cost (C).

   **b.** Now write an equation stating that the total cost is $48.

   How many hats and how many whistles could Randy have bought?

   **c.** Find two other solutions to your equation.

7. Recall the formula tying together distance ($d$), constant speed ($v$), and time ($t$): $\boldsymbol{d = vt.}$
   Sarah jogs at the speed of 9 km per hour, and she rides her bicycle at the speed of 18 km per hour.

   **a.** Convert these speeds to kilometres per minute.

   **b.** Write an expression for the total distance ($d$) Sarah covers
   in $x$ minutes of jogging plus $y$ minutes of bicycling.

   **c.** What distance does Sarah cover if she jogs for
   20 minutes and bicycles for 10 minutes?

   **d.** Let's say the distance Sarah covers, jogging
   and bicycling, is 30 km. Write an equation
   stating this. How many minutes could she
   have jogged/bicycled? Find three possible
   solutions.

   **e.** Write the equation in slope-intercept form and
   plot it.

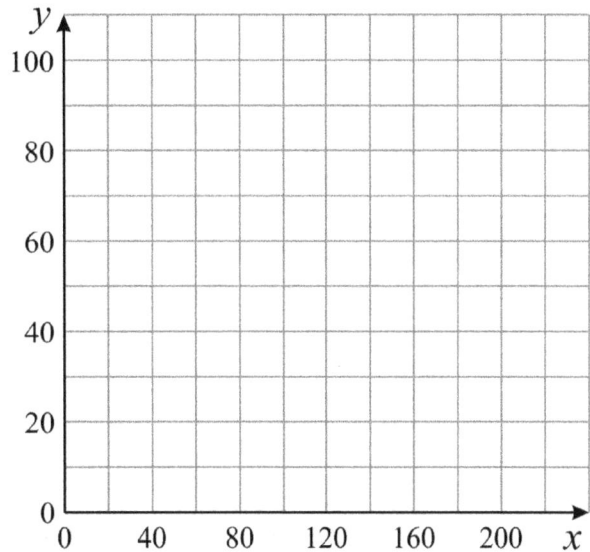

8. General admission to a gardening seminar was $15 but seniors paid only $10. If the total of the admission
   fees was $900, give three possibilities as to how many non-seniors and how many seniors could have attended.

9. A mystery basket contains a mixture of adult cats and kittens (it could even contain zero adults or zero kittens).
Each cat weighs 4 kg and each kitten weighs 0.5 kg.
The total weight of the cats and kittens is 20 kg.

   **a.** If there are $x$ cats and $y$ kittens, write an equation to match the situation.

   **b.** How many adult cats and how many kittens could there be? Find at least three different solutions.

   **c.** Plot your equation from (a).

   **d.** If $x = 1.5$, what is $y$?
   Why is this not a valid solution?

   Plot the individual points on the graph that *are* valid solutions.

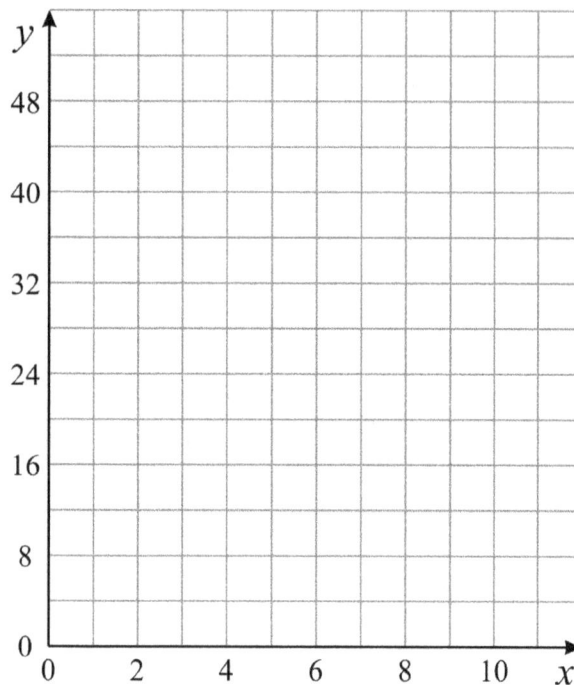

10. Ava and her family went to stay in a resort for a few nights. Each night cost $120 (for the whole family). The resort offered horse rides for $20 per person.

    **a.** If the family stayed for $x$ nights and did $y$ horse rides in total, write an expression for the total cost of these two things.

    **b.** In total, Ava's family spent $760 on the horse rides plus the nights they stayed. How many nights and how many horse rides could they have paid for?

11. The equation $2x^2 - 6x - y = 5$ is a quadratic equation because the variable $x$ is squared. If $x = 0$, then $y = -5$, so $(0, -5)$ is one solution to the equation. Find two other solutions to it.

# Solving Systems of Equations by Graphing

**A system of equations** consists of several equations that have the same variables.

A **solution** to a system of equations is a list of values of the variables that satisfy *all* the equations in the system. For two equations, this is an ordered pair.

**Example 1.** This system of equations consists of two equations. We signify the system with a bracket.

$$\begin{cases} 5x + 4y = 12 \\ y = -x + 2 \end{cases}$$

The solution to the above system is the ordered pair $(4, -2)$, because those values make both equations true: $5(4) + 4(-2)$ does equal 12, and $-2$ does equal $-4 + 2$.

**Example 2.** The equation $y = (3/2)x - 4$ has an infinite number of solutions, and we can represent those solutions with a line drawn in the coordinate plane.

Similarly, the equation $y = -2x + 3$ has infinitely many solutions.

Here is a system of equations consisting of both:

$$\begin{cases} y = (3/2)x - 4 \\ y = -2x + 3 \end{cases}$$

Since the solutions to the first equation form a line, and the solutions to the second also form a line, what would the point of intersection $(2, -1)$ signify?

_____

_____

(The answer is found at the end of the lesson.)

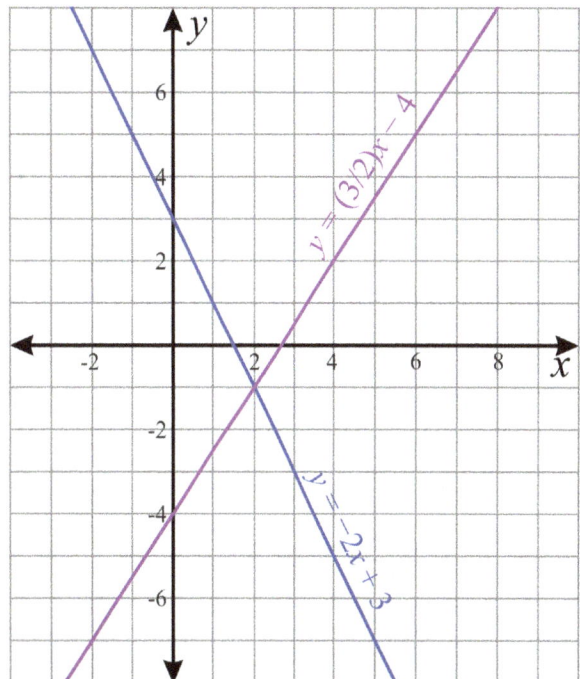

1. Solve each system of equations using the image. The lines are already plotted in it.

   **a.** $\begin{cases} y = -7x - 23 \\ y = (1/3)x - 1 \end{cases}$

   Solution: ( \_\_\_\_\_ , \_\_\_\_\_ )

   **b.** $\begin{cases} y = -(1/4)x + 4 \\ y = -2x + 6 \end{cases}$

   Solution: ( \_\_\_\_\_ , \_\_\_\_\_ )

   **c.** $\begin{cases} -(1/3)x + y = -1 \\ 2x + y = 6 \end{cases}$

   Solution: ( \_\_\_\_\_ , \_\_\_\_\_ )

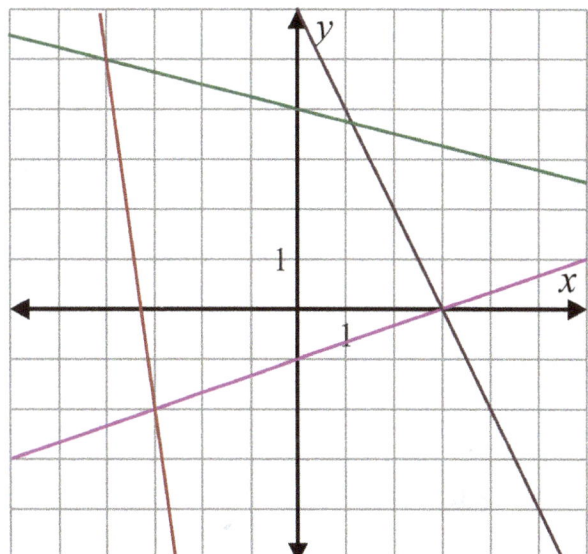

2. Solve each system of equations by graphing.

a. $\begin{cases} y = 2x - 2 \\ y = (2/3)x + 2 \end{cases}$

b. $\begin{cases} 2x - 3y = -6 \\ x + 3y = -3 \end{cases}$

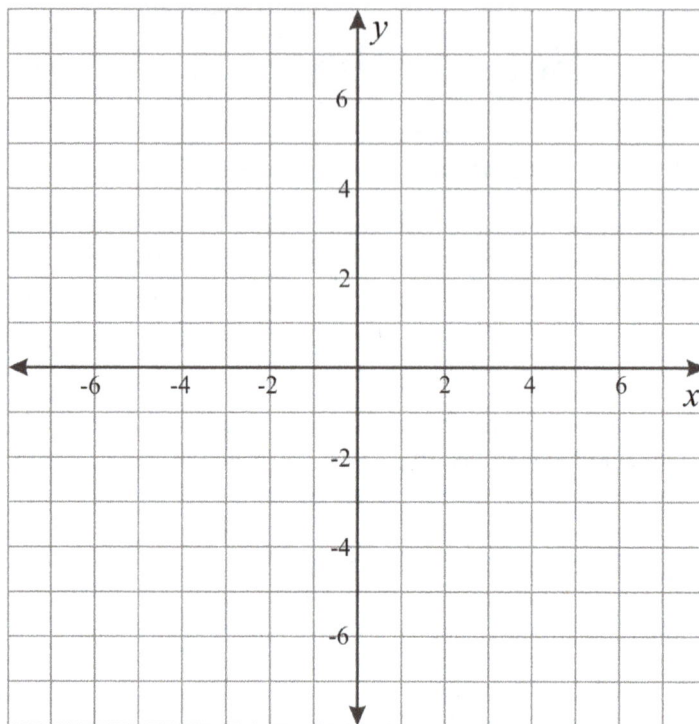

3. Solve each system of equations by graphing. This time, the solutions may not be integers. Give the coordinates of the solutions to the nearest tenth, the best you can by estimating.

a. $\begin{cases} y = -x - 1/2 \\ y = (4/3)x + 1 \end{cases}$

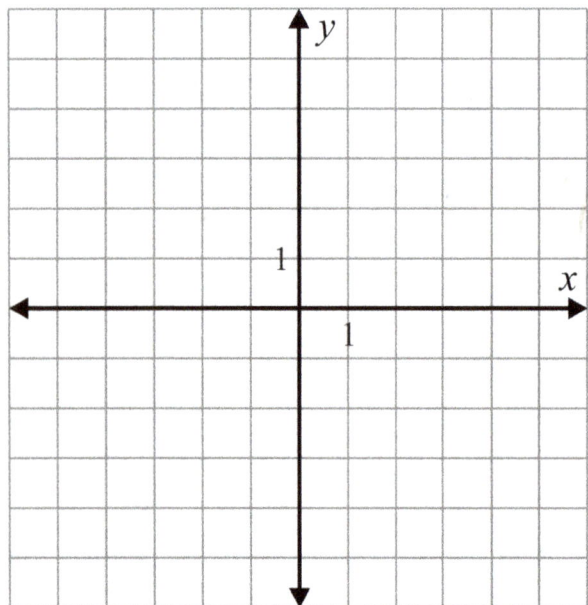

b. $\begin{cases} y = (3/5)x - 4 \\ 3x - 2y - 12 = 0 \end{cases}$

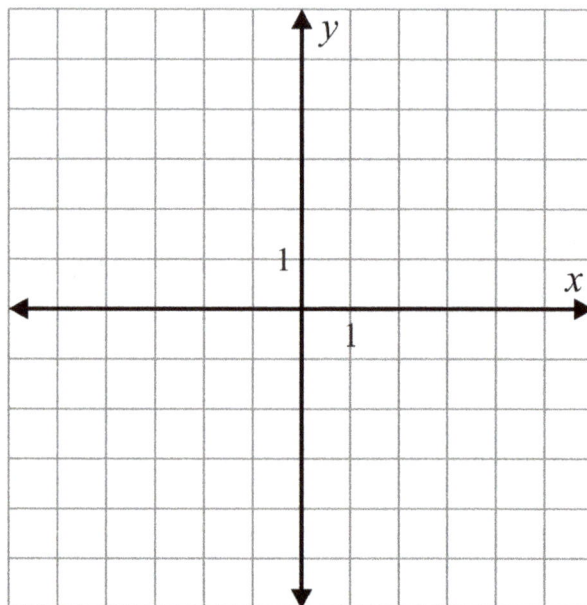

4. (optional) Use Desmos (https://www.desmos.com/calculator) or some other graphing calculator. Plot the lines from the previous exercise. Zoom in as necessary, to find the coordinates of the the point of intersection to the nearest tenth, and thus check your own work for question #3.

5. Party hats cost $2 apiece and party whistles $3 apiece. Randy buys *x* hats and *y* whistles for a total of $24.

   **a.** Write an equation for his total cost in terms of *x* and *y*, and plot it.

   **b.** Let's say that Randy bought a total of 11 hats and whistles.
   Write this as an equation, also, and plot it.

   **c.** How many hats and how many whistles did Randy get?

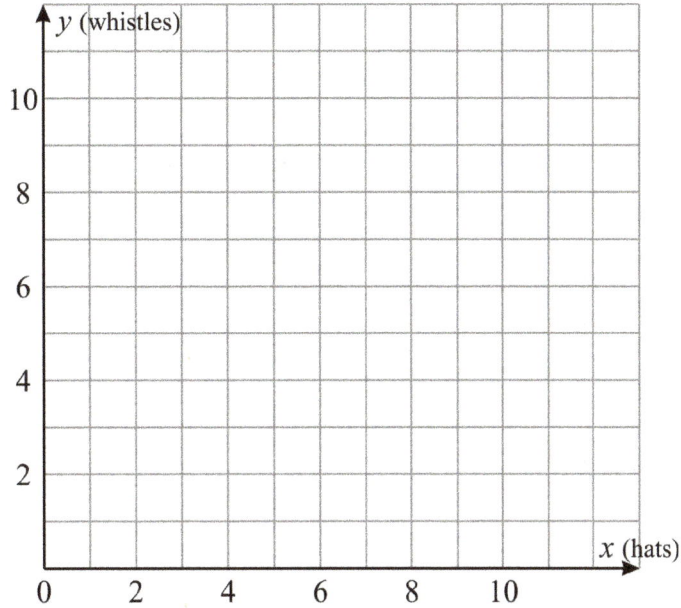

6. Write a system of equations represented by the graphs, and find its solution. Write each linear equation in standard form.

**a.**

$\left\{ \begin{array}{c} \\ \\ \end{array} \right.$

Solution: ( _____ , _____ )

**b.**

$\left\{ \begin{array}{c} \\ \\ \end{array} \right.$

Solution: ( _____ , _____ )

7. A mystery basket contains a mixture of adult cats and kittens (it could even contain zero adults or zero kittens). Each cat weighs 4 kg and each kitten weighs 0.5 kg. The total weight of the cats and kittens is 22 kg.

**a.** If there are $x$ cats and $y$ kittens, write an equation to match the situation, and plot it.

**b.** Let's say there are a total of 16 cats and kittens. Write this as an equation, also, and plot it.

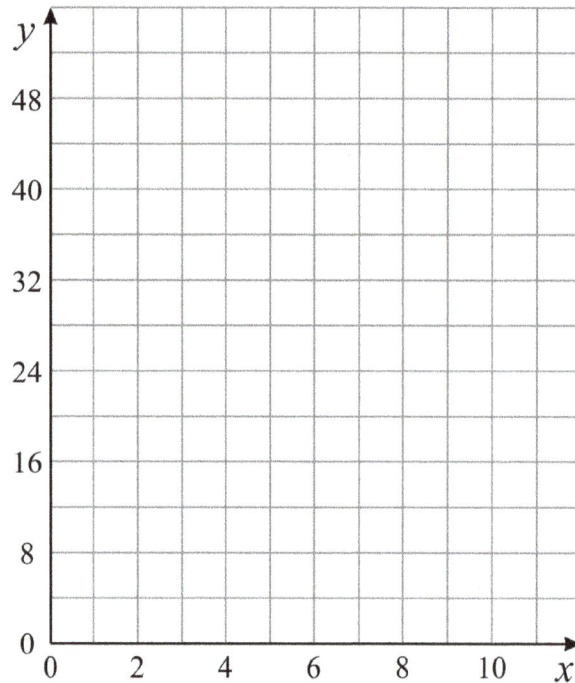

**c.** How many cats and how many kittens are there in the secret basket?

8. **a.** Let's say we have two *distinct* lines that are *not* parallel. How many solutions are there for a system of equations formed from the equations of those two lines?

**b.** Let's say we have two distinct lines that *are* parallel. How many solutions are there for a system of equations formed from the equations of those two lines?

**c.** The third possibility is that we have two identical lines, such as $y = 2x + 1$ and $2x - y = -1$. How many solutions are there for a system of equations formed from those equations?

Find the value of $m$ so that this system will have $(-1, 0)$ as a solution.
$$\begin{cases} y = 2x + 2 \\ y = mx - 6 \end{cases}$$

Puzzle Corner

*Use these problems for extra practice as needed.*

9. Solve each system of equations by graphing.

a. $\begin{cases} y = -x + 5 \\ y = (1/2)x - 4 \end{cases}$

b. $\begin{cases} 3x - 4y = -12 \\ x + 4y = 28 \end{cases}$

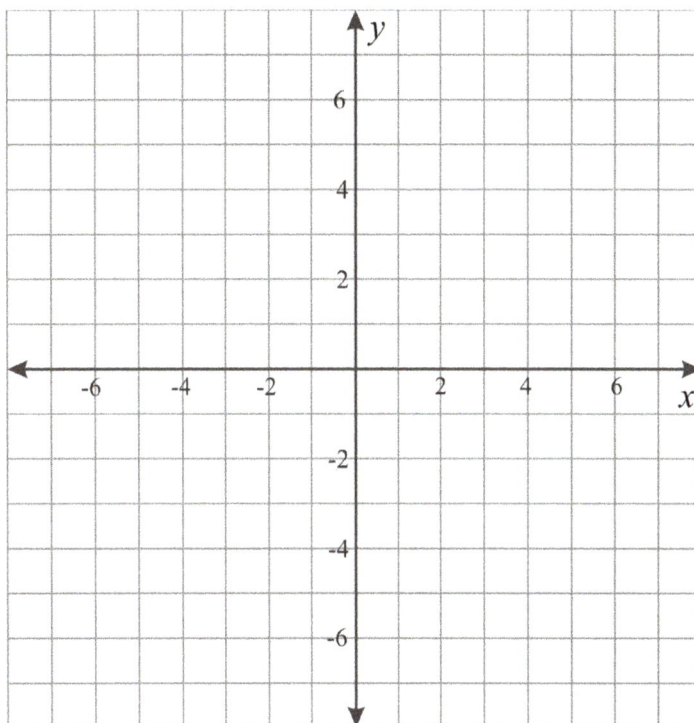

10. Solve each system of equations by graphing. Give the coordinates of the solutions to the nearest tenth, the best you can by estimating.

a. $\begin{cases} y = (1/2)x - 3 \\ 5x + y = -3 \end{cases}$

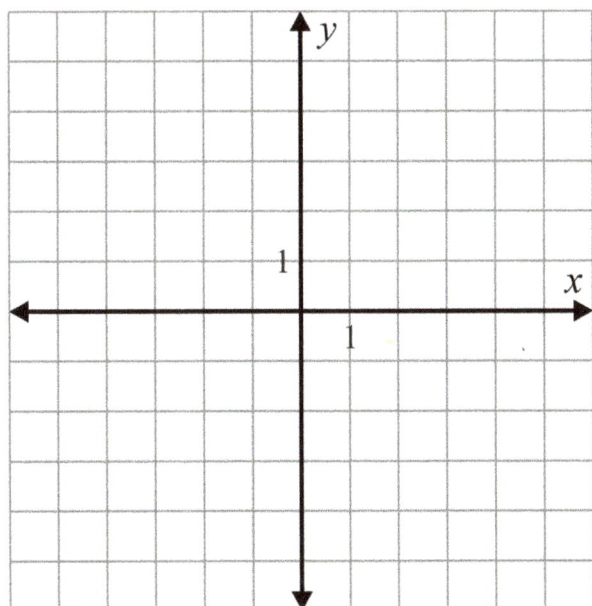

b. $\begin{cases} -2y = 7 - x \\ 4x + y = 2 \end{cases}$

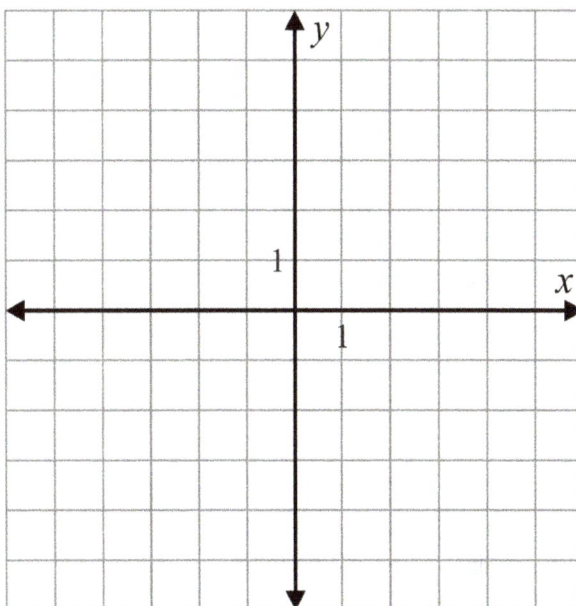

Answer to the question in Example 2. The point of intersection of the two lines is the solution to the system of equations.

# Number of Solutions

1. Solve each system of equations below by graphing. Any surprises?

**a.** $\begin{cases} 2y = 4x - 1 \\ -2x + y = 3 \end{cases}$

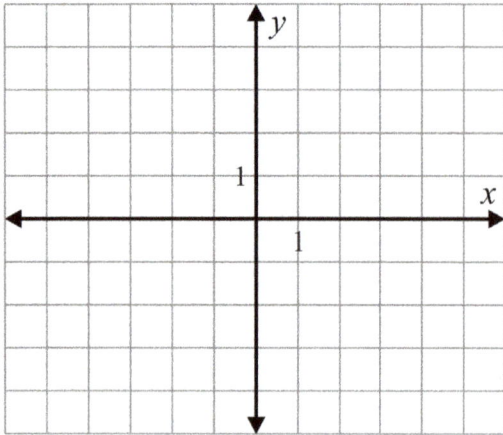

**b.** $\begin{cases} y = -2x + 2 \\ y = x - 4 \end{cases}$

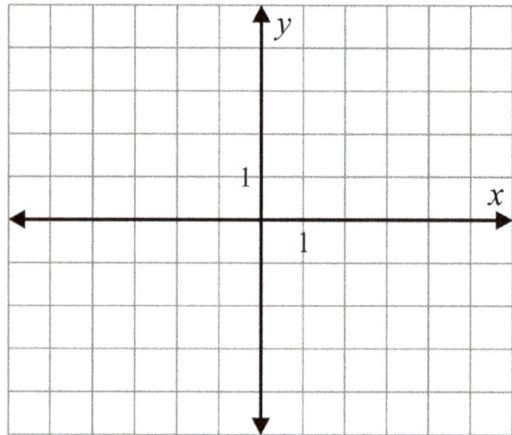

**c.** $\begin{cases} y = -(1/2)x + 1 \\ x + 2y = 2 \end{cases}$

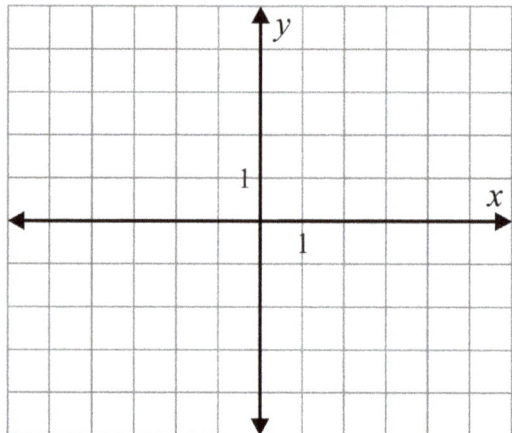

2. **a.** Consider the system of equations on the right. Thinking of the equations as lines in the coordinate system, they both have the **same slope**, 2, but **different y-intercepts**.

   How many solutions does the system have?
   (Does it have any?)
   Explain your thinking.

   $\begin{cases} y = 2x + 3 \\ y = 2x - 6 \end{cases}$

   **b.** Consider the system of equations on the right. Thinking of the equations as lines in the coordinate system, they have a **different slope**.

   How many solutions does the system have?

   How can you tell?

   $\begin{cases} y = -x - 2 \\ y = 3x + 4 \end{cases}$

How many solutions can a system of two linear equations have? If you think about the system graphically, as two lines, and whether they intersect, it is easy to see that there are **three different cases**:

1. The two lines intersect in a single point. Therefore, the system has **one** solution (one ordered pair).
2. The two lines are _____, and thus do not intersect at all. In this case, the system has _____ solutions.
3. The two lines are identical. Here, the system has an _____ number of solutions. It is as if the two lines "intersect in every point". The solution to the system are all points $(x, y)$ that are on the line (satisfy the equation of the line).

Check with your teacher or with the answer key that your answers above are correct.

**Example 1.** This system of equations has an infinite number of solutions. After transforming the top one to the slope-intercept form, we see that both equations are actually the same:

$$\begin{cases} 10x + 5y = -15 \\ y = -2x - 3 \end{cases} \rightarrow \begin{cases} 5y = -10x - 15 \;\big| \div 5 \\ y = -2x - 3 \end{cases} \rightarrow \begin{cases} y = -2x - 3 \\ y = -2x - 3 \end{cases}$$

We cannot list all the solutions, but we can still give the solution, in this manner:

<u>Solution:</u> All points $(x, y)$ that satisfy the equation $y = -2x - 3$. (You can state either of the original equations.)

3. Match each system of equations with its solution.

a. $\begin{cases} y = 20x - 1 \\ y = 5x + 5 \end{cases}$
b. $\begin{cases} y = -(1/3)x - 1 \\ y = -(1/3)x + 1 \end{cases}$
c. $\begin{cases} y = 4x + 1 \\ 2y = 8x + 2 \end{cases}$

All points $(x, y)$ that satisfy the equation $y = 4x + 1$.

There are no solutions.

$(2/5, 7)$

4. Specify the solution for each system of equations represented by the lines, or state that there are no solutions.

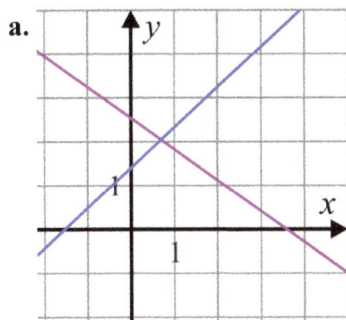

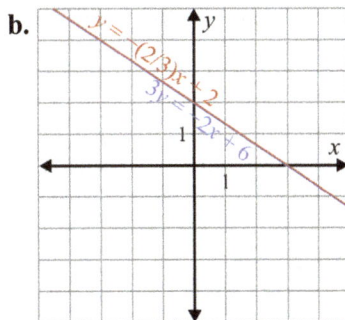

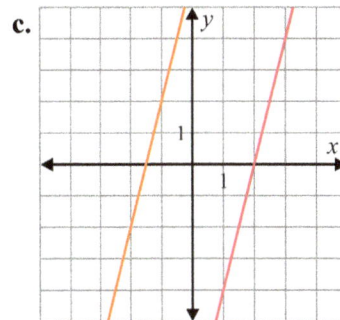

a.
b.
c.

5. Classify each system of equations as having either 0, 1, or an infinite number of solutions. Think of slope, and how it relates to the intersection points of the lines. You may need to transform one or both equations.

a. $\begin{cases} x + y = 15 \\ 3x + y = 15 \end{cases}$
b. $\begin{cases} y = -(1/4)x \\ 4y = -x \end{cases}$
c. $\begin{cases} y = -4x - 1 \\ 4x + y = 0 \end{cases}$

Often, a simple **inspection** of the equations will reveal how many solutions a system of equations has.

**Example 1.** In this system of equations, the expression $2x + y$ is set to equal two different values.

$$\begin{cases} 2x + y = 12 \\ 2x + y = -5 \end{cases}$$

This would mean that $-5 = 12$, which clearly is untrue! There are no values of $x$ and $y$ that can fulfill both equations.

Graphically, these two lines have the same slope, but different $y$-intercepts (check this!), so they are parallel and never meet.

**Example 2.** Here, the second equation can be obtained from the first by multiplying each side by 2.

$$\begin{cases} 3x - 2y = 6 \\ 6x - 4y = 12 \end{cases}$$

If we write each equation in slope-intercept form, they both become the same equation, $y = (3/2)x - 3$. Graphically, both equations represent the same line.

Therefore, the solution to this system is all points $(x, y)$ that satisfy the equation $3x - 2y = 6$ (you could list either equation here). Thus, the system has an infinite number of solutions.

This will always be the case if one equation is obtained from the other by multiplying or dividing both sides by something.

6. Tell how many solutions each system of equations has by inspecting and/or checking the slope and $y$-intercept. You do not have to find the solution(s).

a. $\begin{cases} 5y - 20x = 10 \\ y - 4x = 2 \end{cases}$

b. $\begin{cases} x = y + 2 \\ x = y + 3 \end{cases}$

c. $\begin{cases} x = y + 1 \\ 2x = y + 3 \end{cases}$

d. $\begin{cases} x - 4y = 0 \\ x - 4y = 5 \end{cases}$

e. $\begin{cases} x + y = -1/4 \\ 2y - x = 5 \end{cases}$

f. $\begin{cases} y = -4x - 1 \\ 4x + y = 0 \end{cases}$

g. $\begin{cases} 50x + 20y = 200 \\ 50x - 20y = 300 \end{cases}$

h. $\begin{cases} 60x + 10y = -100 \\ -40y - 240x = 400 \end{cases}$

i. $\begin{cases} 20 - 20x = 10y \\ 20x + 10y = -70 \end{cases}$

7. One of the systems of equations below has an infinite number of solutions. Find that system, and specify what its solution is.

(i) $\begin{cases} 3x + y = -6 \\ y = -3x + 5 \end{cases}$

(ii) $\begin{cases} y = 2x + 5 \\ x + 3y = -6 \end{cases}$

(iii) $\begin{cases} x/3 - y = 2 \\ x = 3y + 6 \end{cases}$

8. Create a system of equations using $y = 3x + 2$ as one of the equations and by adding a second, different equation to the system, in such a manner that...

**a.** the system has no solutions

$$\begin{cases} y = 3x + 2 \\ \end{cases}$$

**b.** the system has one solution

$$\begin{cases} y = 3x + 2 \\ \end{cases}$$

**c.** the system has an infinite number of solutions

$$\begin{cases} y = 3x + 2 \\ \end{cases}$$

9. Use the lines in this graph to build a system of two equations that...

**a.** has $(2, -3)$ as a solution

**b.** has no solutions

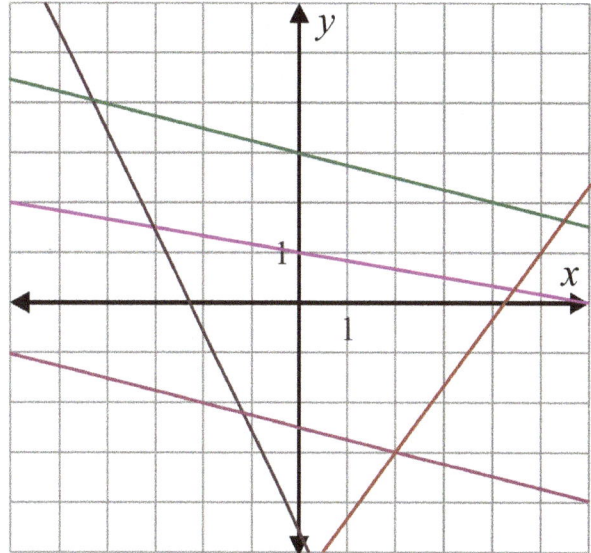

10. Tell how many solutions each system of equations has. You do not have to find the solution(s).

**a.** $\begin{cases} 2y - x = -4 \\ y - 4x = 2 \end{cases}$

**b.** $\begin{cases} -5x = y \\ -5x - y = 3 \end{cases}$

**c.** $\begin{cases} x - y = -1 \\ x - y = 3 \end{cases}$

**d.** $\begin{cases} -2x + y = 1 \\ 2x - y = -1 \end{cases}$

**e.** $\begin{cases} x/2 + y/2 = -1 \\ x - y = -2 \end{cases}$

**f.** $\begin{cases} x/2 - y/2 = -1 \\ x - y = -2 \end{cases}$

---

Solve System 2, knowing that System 1 has no solutions.

**Puzzle Corner**

System 1:
$$\begin{cases} 4x - y = 0 \\ y = sx + 5 \end{cases}$$

System 2:
$$\begin{cases} x - 2 = 0 \\ y = 3x + s \end{cases}$$

# Solving Systems of Equations by Substitution

While graphing is a valid way to solve systems of equations, it is not the best since the coordinates of the intersection point may be decimal numbers, and even irrational. In this lesson you will learn one algebraic method for solving systems of equations, called **the substitution method**.

**Example 1.** Note that the second equation in this system of equations is of the form "$y$ = something", and this "something" only involves the variable $x$.

$$\begin{cases} 5x - 2y = 16 \\ y = -2x + 1 \end{cases}$$

This means we can replace $y$ in the *first* equation by the expression that $y$ equals (which is $-2x + 1$), and the result will be an equation that will have **only one unknown**, $x$.

We can then solve that equation for $x$. Once we know the value of $x$, we can substitute that value in one of the original equations, and thus get an equation only in $y$. Solving that, we will find the value of $y$.

So, let's do all that!
Since $y = -2x + 1$, we will substitute $-2x + 1$ in place of $y$ in the first equation. →

$$\begin{aligned} 5x - 2y &= 16 \\ 5x - 2(-2x + 1) &= 16 \\ 5x + 4x - 2 &= 16 \\ 9x - 2 &= 16 \\ 9x &= 18 \\ x &= 2 \end{aligned}$$

Once we solve that $x = 2$, we then substitute $2$ in place of $x$ in the second equation:

$$y = -2x + 1 = -2(2) + 1 = -3$$

So, the solution is $x = 2$ and $y = -3$. Let's check:

$$\begin{cases} 5(2) - 2(-3) \overset{?}{=} 16 \\ -3 \overset{?}{=} -2(2) + 1 \end{cases} \rightarrow \begin{cases} 10 + 6 = 16 \checkmark \\ -3 = -4 + 1 \checkmark \end{cases}$$

1. Solve each system of equations using the substitution method. Check your solutions.

**a.** $\begin{cases} x + y = -9 \\ y = 3x - 1 \end{cases}$

**b.** $\begin{cases} x = 3y - 11 \\ 2x + 2y = 10 \end{cases}$

**Example 2.** Here, we have added line numbers to the equations, just to be able to reference each equation.

$$\begin{cases} 6x - 2y = -12 & (1) \\ x + 3y = -10 & (2) \end{cases}$$

This time, neither equation is in the form "$y$ = something" or "$x$ = something", so we cannot substitute any expression directly to either equation. But, we can solve either $x$ or $y$ from either equation, and then use the substitution method.

$$(2) \quad x + 3y = -10$$
$$x = -3y - 10$$

Let's solve $x$ from equation 2, which is relatively simple to do. We mark (2) in front of it to signify it is equation 2 we are using.

Now, we substitute the expression **$-3y - 10$** in place of $x$ in equation 1, resulting in an equation that only has one unknown.

$$(1) \qquad 6x - 2y = -12$$
$$6(-3y - 10) - 2y = -12$$
$$-18y - 60 - 2y = -12$$
$$-20y - 60 = -12 \quad \big| + 60$$
$$-20y = 48 \quad \big| \div (-20)$$

Then we solve this equation for $y$.

$$y = -48/20 = -12/5$$

Then it's time to substitute this value $-12/5$ in place of $y$ in *either* of the original equations, to find the value of $x$. Let's use equation 2, since in it, $x$ does not have any coefficients so the calculation will be shorter.

$$(2) \quad x + 3y = -10$$
$$x = -3y - 10$$
$$x = -3(-12/5) - 10$$
$$x = 36/5 - 10 = 36/5 - 50/5 = -14/5$$

The solution is $(-14/5, -12/5)$.

---

2. Solve each system of equations using the substitution method. Check your solutions (always!).

| a. $\begin{cases} x - 4y = -2 \\ y = 5 - 2x \end{cases}$ | b. $\begin{cases} x = 10y + 1 \\ (1/2)x - y = 3 \end{cases}$ | c. $\begin{cases} 3x = 3(y - 1) \\ y = 5x \end{cases}$ |
|---|---|---|
|  |  |  |

3. Find the errors that the students made in solving these systems of equations.

**a.**

$$\begin{cases} x + 3y = -5 & \text{(1)} \\ 2x + y = 3y - 1 & \text{(2)} \end{cases}$$

$\downarrow$

$$\begin{cases} x = -3y - 5 & \text{(1)} \\ 2x + y = 3y - 1 & \text{(2)} \end{cases}$$

$\downarrow$

(1) $\quad (\mathbf{-3y - 5}) + 3y = -5$

$\qquad -3y - 5 + 3y = -5$

$\qquad\qquad\qquad -5 = -5$

*What now? Does this mean*
*y can be any number?*

**b.**

$$\begin{cases} -4x - y = 2(x - 3) & \text{(1)} \\ x + y = 3 & \text{(2)} \end{cases}$$

$\downarrow$

$$\begin{cases} -4x - y = 2(x - 3) & \text{(1)} \\ y = 3 - x & \text{(2)} \end{cases}$$

$\downarrow$

(1) $\quad -4x - (\mathbf{3 - x}) = 2(x - 3)$

$\qquad\qquad -3x - 3 = 2x - 6$

$\qquad\qquad -5x - 3 = -6$

$\qquad\qquad\qquad -5x = -3$

$\qquad\qquad\qquad\quad x = 3/5$

The solution is 3/5.

4. Find the error in this solution, also.

$$\begin{cases} 5x - 2y = 10 & \text{(1)} \\ 3x - 8y = 4y - 48 & \text{(2)} \end{cases}$$

$\downarrow$

$$\begin{cases} 5x = 2y + 10 & \text{(1)} \\ 3x - 8y = 4y - 48 & \text{(2)} \end{cases}$$

$\downarrow$

(2) $\quad 3(\mathbf{2y + 10}) - 8y = 4y - 48$

$\qquad\quad 6y + 30 - 8y = 4y - 48$

$\qquad\qquad -2y + 30 = 4y - 48$

$\qquad\qquad\qquad -6y = -78$

$\qquad\qquad\qquad\quad y = 13$

Substituting $y = 13$ to the first equation:

(1) $\quad 5x - 2(13) = 10$

$\qquad\quad 5x - 26 = 10$

$\qquad\qquad\quad 5x = 36$

$\qquad\qquad\quad\ x = 7.2$

(*Continues in the column on the right.*)

However, (7.2, 13) does NOT fulfill the
second equation:

(2) $\quad 3x - 8y = 4y - 48$

$3(7.2) - 8(13) \overset{?}{=} 4(13) - 48$

$\qquad 21.6 - 104 \neq 4$

What happens if a system of equations has no solutions, or an infinite number of solutions? What will that look like in an algebraic solution?

**Example 3.** The system below actually has no solutions. Let's see what happens when we try to solve it.

$$\begin{cases} 8x + y = 3(1 + y) & (1) \\ -4x + y = -5 & (2) \end{cases}$$

First we solve $y$ from the second equation:

$$(2) \quad -4x + y = -5$$
$$y = 4x - 5$$

Now we substitute that for $y$ in the first equation:

$$(1) \quad 8x + (4x - 5) = 3[1 + (4x - 5)]$$
$$12x - 5 = 3(1 + 4x - 5)$$
$$12x - 5 = 3(4x - 4)$$
$$12x - 5 = 12x - 12$$
$$-5 = -12$$

We end up with a false equation. That is how we can see that the system has no solutions.

**Example 4.** $\begin{cases} 3x - y = 2(x - 3) & (1) \\ x + 5y = 6(y - 1) & (2) \end{cases}$

First we solve $y$ from the first equation:

$$(1) \quad 3x - y = 2(x - 3)$$
$$-y = 2(x - 3) - 3x$$
$$-y = 2x - 6 - 3x$$
$$y = x + 6$$

Now we substitute that for $y$ in the second equation:

$$(2) \quad x + 5(x + 6) = 6[(x + 6) - 1]$$
$$x + 5x + 30 = 6[x + 6 - 1]$$
$$6x + 30 = 6x + 30$$

Right here we can see that any $x$ satisfies the above equation. Or, we could continue by subtracting 30 and $6x$ from both sides and arrive at the identity $0 = 0$.

Arriving at this situation means that the original two equations are equivalent (the two lines are the same line). However, it doesn't mean both $x$ and $y$ can be just anything. The pair $(x, y)$ has to be a point on this line.

The solution needs to include the equation that $x$ and $y$ satisfy (the equation of the line). It can be *either* of the original equations, or an equivalent equation. Here it makes sense to use $y = x + 6$ since it is so simple.

Solution: All points $(x, y)$ that satisfy the equation $y = x + 6$.

5. Solve each system of equations using the substitution method. Use extra paper if necessary.

a. $\begin{cases} 2x = 14 - 2y \\ x + y = 2 \end{cases}$

b. $\begin{cases} 2 - 4x = y + 8 \\ x - y = 5x + 6 \end{cases}$

6. Solve each system of equations. Use extra paper if necessary. Check your solutions (always do that!).

a. $\begin{cases} 4x - 2y = 5 \\ y = 5x + 8 \end{cases}$

b. $\begin{cases} 4x - 5y = 13 \\ y = 2(5 - x) \end{cases}$

c. $\begin{cases} x - 2y = 5 - 3.5y \\ 2(5 - x) = 3y \end{cases}$

d. $\begin{cases} -4x - 3y = 2 \\ y + 4x = -6 \end{cases}$

e. $\begin{cases} -9x = -2y \\ 9x - 2y = 2 \end{cases}$

f. $\begin{cases} 2(2x - y) = 2 \\ x + 6y = 1 \end{cases}$

7. Solve each system of equations by substitution. Then graph the lines. Verify that the intersection point of the lines is the solution you found algebraically.

a. $\begin{cases} 2x - 5y = 20 \\ -x - y = 3 \end{cases}$

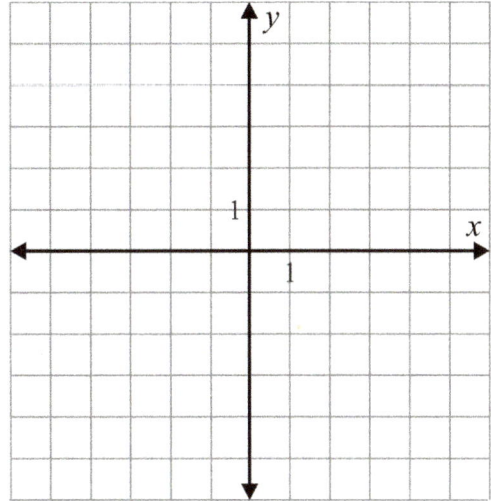

b. $\begin{cases} -x + 5y = 20 \\ y = 3x \end{cases}$

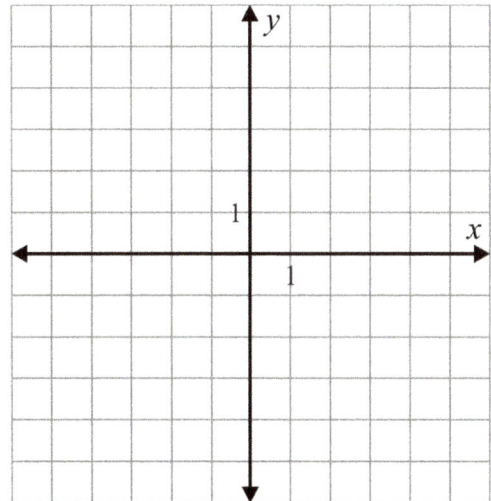

8. Solve each system of equations. Round the coordinates to two decimal digits.

a. $\begin{cases} 0.2x - 0.7y = 1 \\ x + y = 4 \end{cases}$

b. $\begin{cases} 1.2x - 3y = -2 \\ y = -0.6x + 5 \end{cases}$

9. These are for additional practice as needed. Use extra paper if necessary, and check your solutions.

a. $\begin{cases} x = -(y - 2) \\ 3y = 10(2 - x) \end{cases}$

b. $\begin{cases} 4x - 3y = -1 \\ 3x + y = -2 \end{cases}$

c. $\begin{cases} x = -3(y + 8) \\ -x + 3y = 0 \end{cases}$

d. $\begin{cases} 3(x - 2) + y = 12 \\ y = 4x - 3 \end{cases}$

e. $\begin{cases} 2x - 8y = 15 \\ -x + y = 20 \end{cases}$

f. $\begin{cases} x - 3(3y + 2) = 1 \\ 3x + 5y = -1 \end{cases}$

# Applications, Part 1

**Example 1.** Sandra gave her kids a riddle to solve: "I have a total of 38 animals — some are kittens and some are chicks. They have a total of 98 legs. How many kittens and how many chicks are there?"

This type of problem is easily solved by using two variables, and writing two equations from the given information.

Let $x$ be the number of kittens and $y$ be the number of chicks. Then, we know that $x + y = 38$.
Since each kitten has 4 legs, the kittens have a total of $4x$ legs. Similarly, the chicks have $2y$ legs.

So, the other equation we get is $4x + 2y = 98$.

$$(1) \quad \begin{cases} x + y = 38 \\ 4x + 2y = 98 \end{cases} (2)$$

Now let's solve our system of equations!

Since the top equation has $y$ by itself, we will use the substitution method, and solve $y$ from the top equation.

$$(1) \quad x + y = 38$$
$$y = 38 - x$$

Then we substitute $38 - x$ in place of $y$ in the second equation, and get that $x = 11$. In other words, there are 11 kittens.

Then it's easy to find out that there are $38 - 11 = 27$ chicks.

This checks, because $4 \cdot 11 + 2 \cdot 27 = 44 + 54 = 98$ legs.

$$\begin{aligned} (2) \qquad 4x + 2y &= 98 \\ 4x + 2(38 - x) &= 98 \\ 4x + 76 - 2x &= 98 \\ 2x + 76 &= 98 \quad \Big| -76 \\ 2x &= 22 \quad \Big| \div 2 \\ x &= 11 \end{aligned}$$

1. An amusement park sells tickets for children for \$45 and for adults for \$70. When a group of 15 people (some adults, some children) went to the park, the total cost was \$900. How many adults and how many children were in the group?

2. Ice skating at a local skating rink costs $5 per hour, but weekends are costlier, $7 per hour.
Last February, Suzie skated a total of 26 hours, and her total cost was $142.
How many hours did she skate during weekends?

3. A piggy bank has a total of 76 coins, in dimes and nickels, and
they're worth $5.00. Jamie wrote the following equations for this problem:

Are they correct? If not, correct one or both of them. Then find out how many
dimes and how many nickels there are.

$$d + n = 76$$

$$10d + 5n = 5$$

4. Lily's piggy bank has only quarters and dimes. The
bank contains 44 coins and they're worth $6.65.
How many quarters and dimes does she have?

5. In a barnyard, 57 animals just went by, and they had 158 legs. If the animals consisted of sheep and ducks, how many sheep and how many ducks were there?

6. There are two routes that lead from Michael's house to his grandmother's place. The longer route is 9 km shorter than double the shorter route. The shorter route is 4/5 as long as the longer. Find the lengths of the two routes.

7. Ava and Eva were comparing their ages. Ava said, "My age is double your age minus 15 years." Eva said, "And my age is 6/7 of your age." Find their ages.

8. Juan said to Henry, "In ten years, my age will be double your age." And Henry said, "Five years ago, your age was five times mine." Find their ages now.

9. Amanda sells home-made fruit bars for $6 each and nut bars for $10 each. One day, she sold five more nut bars than fruit bars, and her total sales were $242. How many fruit bars did she sell that day?

Puzzle Corner Diane gave her friends a riddle to solve: "I have a total of $3.25 in 25, 10, and 5 cent coins. There are twice as many dimes as quarters, and 10 more nickels than dimes."
Find the quantity of each coin Diane has.

# The Addition Method, Part 1

The addition method for solving a system of linear equations is based on this principle. If we have two things that equal each other, such as A = B, and *other* two things that also are equal, such as C = D, then if we add A and C, that equals B + D.

We can write the two equations one under the other, then draw a line like when adding in columns. Then when we add the left sides and the right sides, the equality is preserved:

$$\begin{array}{rcl} A & = & B \\ + \quad C & = & D \\ \hline A + C & = & B + D \end{array}$$

Here is a numerical example. Verify that the last line is a true equation.

Adding the equations works well for certain systems of equations. See the examples in the lesson.

$$\begin{array}{rcl} 2 + 5 & = & 7 \\ + \quad\quad 4 & = & 10 - 6 \\ \hline 2 + 5 + 4 & = & 7 + 10 - 6 \end{array}$$

**Example 1.** Note that in this system of equations, the first equation has the term $-y$ and the second has $y$.

When we add the equations, those two terms will be *eliminated*, and the sum will be an equation with only one variable, $x$. This is why the addition method is also called **the elimination method**.

We can easily solve $x$ from the resulting equation to get $x = 2$.

$$+ \left\{ \begin{array}{rcll} 5x - y & = & 4 & \text{(1)} \\ 7x + y & = & 20 & \text{(2)} \\ \hline 12x & = & 24 & \\ x & = & 2 & \end{array} \right.$$

Then we substitute that value of $x$ to *either* of the original equations, (which then becomes an equation in only $y$), and solve for $y$:

We get that $y = 6$. So, the solution is the ordered pair (**2**, **6**). Let's check:

$$\begin{array}{rcll} 5(2) - y & = & 4 & \text{(1)} \\ 10 - y & = & 4 & \\ y & = & 6 & \end{array}$$

$$\left\{ \begin{array}{l} 5(2) - 6 \overset{?}{=} 4 \\ 7(2) + 6 \overset{?}{=} 20 \end{array} \right. \quad \rightarrow \quad \left\{ \begin{array}{l} 10 - 6 = 4 \quad \checkmark \\ 14 + 6 = 20 \quad \checkmark \end{array} \right.$$

1. Solve each system of equations using the addition (elimination) method.

**a.** $\left\{ \begin{array}{l} x + y = -7 \\ 3x - y = 3 \end{array} \right.$

**b.** $\left\{ \begin{array}{l} -x + 3y = 11 \\ x + 2y = 9 \end{array} \right.$

**Example 2.** In this system, the terms that will get eliminated are $6x$ and $-6x$.

Note that we write the final $x$-value, $17/6$, as a fraction, not as a mixed number. This is because a fraction is less likely to be misread than a mixed number when there are other numbers and symbols around it.

Lastly, here's the check for the solution. That is an important step! The examples and the answer key don't always include it, but you should always do it with your own work.

$$
\begin{cases}
6(17/6) + 3(-4) \overset{?}{=} 5 \\
-6(17/6) - 7(-4) \overset{?}{=} 11
\end{cases}
\rightarrow
\begin{cases}
17 - 12 = 5 \checkmark \\
-17 + 28 = 11 \checkmark
\end{cases}
$$

$$
\begin{aligned}
& \phantom{+} \begin{cases} 6x + 3y = 5 \quad (1) \\ -6x - 7y = 11 \quad (2) \end{cases} \\
& \overline{\phantom{+ \{} \quad -4y = 16} \\
& \phantom{+ \{ -4} \quad y = -4
\end{aligned}
$$

$$
\begin{aligned}
6x + 3(-4) &= 5 \quad (1) \\
6x - 12 &= 5 \\
6x &= 17 \\
x &= 17/6
\end{aligned}
$$

Solution: $(17/6, -4)$

2. Solve each system of equations using the elimination method. Check your solutions — as always!

a. $\begin{cases} 2x + 2y = 7 \\ -2x - 12y = 23 \end{cases}$

b. $\begin{cases} 8x - 5y = 9 \\ -8x - y = 21 \end{cases}$

c. $\begin{cases} 2x - 3y = 11 \\ -2x + 3y = -11 \end{cases}$

d. $\begin{cases} 4x + 9y = 16 \\ 4x + 2y = 2 \end{cases}$

**Example 3.** Here we need to first transform equation 1 so that both the $x$ and $y$-terms are on the left side. In equation 2, we combine like terms. Only after that do we add the equations.

$$\begin{cases} 7x = y - 5 & (1) \\ x + y + x = 11 & (2) \end{cases} \rightarrow \quad + \begin{cases} 7x - y = -5 & (1) \\ 2x + y = 11 & (2) \end{cases}$$

$$\frac{\phantom{xxxxxxxxxxxxx}}{9x = 6}$$

$$x = 2/3$$

Then we continue as usual, substituting $x = 2/3$ into the second equation, and thus solve for y:

$$2(2/3) + y = 11 \qquad (2)$$
$$4/3 + y = 11$$
$$y = 11 - 4/3$$
$$y = 9\ 2/3 \text{ or } 29/3$$

Solution: (2/3, 29/3)

3. Solve. Use extra paper if necessary. Check your solutions.

**a.**
$$\begin{cases} 7x = 10 - 4y \\ -x - 4y = 26 \end{cases} \rightarrow$$

**b.**
$$\begin{cases} -2x + 6y = 4 \\ 4 + 5x = 2y + 9 + 4y \end{cases} \rightarrow$$

**c.**
$$\begin{cases} 10 = 2x - 4y - 2 \\ -2x + 7y = 18 + 2y \end{cases} \rightarrow$$

**d.**
$$\begin{cases} 50x + 120 = 20y \\ 35y - 50x = 180 \end{cases} \rightarrow$$

**Example 4.** The line diagram on the right has two rows of line segments. The top one is $2x + 2$ long, and the bottom one is $x + 8$ long. Together, they illustrate the equation $2x + 2 = x + 8$.

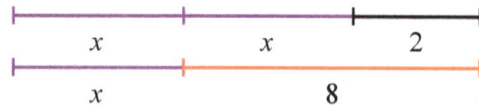

If you ignore one $x$ from both rows, it is easy to see <u>visually</u> that $x + 2$ must equal 8, so $x$ must equal 6.

**Example 5.** The diagram on the right illustrates this system of equations:

$$\begin{cases} 2x + y = 22 \\ 3x + y = 31 \end{cases}$$

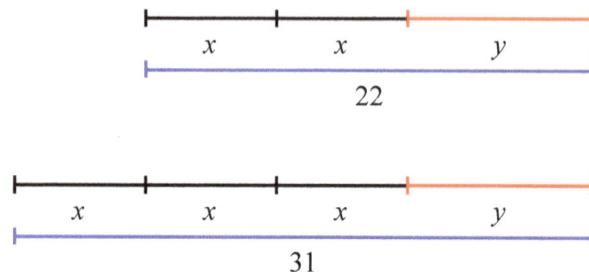

4. Solve the system of equations in example 5 *visually*, using the diagrams. See a hint at the bottom of the page.

5. Solve the systems of equations using the diagrams or algebraically.

a.

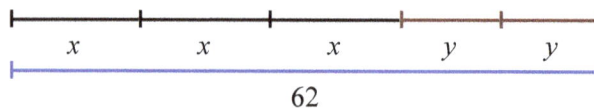

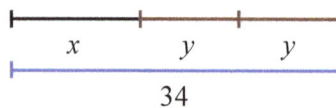

b.

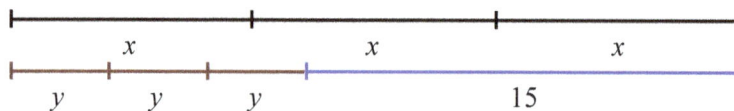

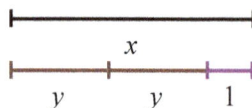

*Hint for problem 4*: note that if $2x + y$ equals 22, then $2x + y$ in the second diagram, top line, is also worth 22, which only leaves one $x$ and 22 being worth 31.

6. Use these problems for extra practice. Use extra paper if necessary.

a. $\begin{cases} 2x - 10y = -36 \\ x + 10y = 24 \end{cases}$

b. $\begin{cases} 5(x + y) = 20 \\ 5x + 4y = 34 \end{cases}$

c. $\begin{cases} 5x - 11 = -2y \\ 5x + 2y = 14 \end{cases}$

d. $\begin{cases} 4(x - 1) = 2y \\ 3y = -4x + 8 \end{cases}$

e. $\begin{cases} 0.2x - 2 = 0.1y \\ 0.1y = -0.2x + 2 \end{cases}$

f. $\begin{cases} 200x - 120y = 200 \\ 120y + 400x = 600 \end{cases}$

# The Addition Method, Part 2

**Example 1.** In this system of equations, it will not work to add the equations. Neither the $x$ nor the $y$ terms have opposite coefficients, and will not cancel out.

$$\begin{cases} 5x - 6y = 12 \\ x - 2y = 4 \end{cases} \quad \Big| \cdot (-5)$$

But we can make it so! If we **multiply the bottom equation** by $-5$, the new set of equations will have $5x$ in the top equation and $-5x$ in the bottom. The $5x$ and $-5x$ are called **a zero pair**, and they will cancel out when we add the equations.

$$+ \begin{cases} 5x - 6y = 12 \quad (1) \\ -5x + 10y = -20 \quad (2) \end{cases}$$
$$4y = -8$$
$$y = -2$$

Doing that, we easily solve that $y = -2$. Then, substituting that value to the first equation, we find out that $x = 0$.

So, the solution is the ordered pair ($0$, $-2$). Let's check that with the *original* equations:

$$5x - 6(-2) = 12 \quad (1)$$
$$5x = 0$$
$$x = 0$$

$$\begin{cases} 5(0) - 6(-2) \overset{?}{=} 12 \\ 0 - 2(-2) \overset{?}{=} 4 \end{cases} \rightarrow \begin{cases} 12 = 12 \ \checkmark \\ 4 = 4 \ \checkmark \end{cases}$$

Why is this okay to do? If you multiply both sides of an equation by a non-zero number, the resulting equation is equivalent to the original: it has the same solution(s).

1. Solve each system of equations. In (a), first multiply the top equation by 2. In (b), you figure out which equation to multiply by what, at first. Then add the equations.

**a.**
$$\begin{cases} 3x + y = -7 \\ -6x - 4y = 8 \end{cases} \quad \Big| \cdot 2$$
$$\downarrow$$
$$+ \begin{cases} \underline{\phantom{2}}x + \underline{\phantom{2}}y = \underline{\phantom{2222}} \\ -6x - 4y = 8 \end{cases}$$

**b.**
$$\begin{cases} -5x + 3y = -8 \\ 3x - y = 4 \end{cases}$$

**Example 2.** In this system of equations, we will transform *both* the top and the bottom equations, in order to arrive at a zero pair.

(1) $\begin{cases} 2x - 4y = 20 \\ 3x + 6y = -6 \end{cases}$ $\begin{array}{l} \cdot\,(-3) \\ \cdot\,2 \end{array}$
(2)

When we multiply the top equation by $-3$ and the bottom equation by $2$, the new set of equations will have $-6x$ in the top equation and $6x$ in the bottom (a zero pair).

Doing that, we easily solve that $y = -3$. Then, substituting that value to the second equation, we find out that $x = 4$.

So, the solution is the ordered pair $(4, -3)$. Let's check that with the *original* equations:

$\begin{cases} 2(4) - 4(-3) \stackrel{?}{=} 20 \\ 3(4) + 6(-3) \stackrel{?}{=} -6 \end{cases}$ $\rightarrow$ $\begin{cases} 8 + 12 = 20 \; \checkmark \\ 12 - 18 = -6 \; \checkmark \end{cases}$

$\downarrow$

$+ \begin{cases} -6x + 12y = -60 \\ \phantom{-}6x + 12y = -12 \end{cases}$
$$\overline{\phantom{+ \begin{cases}-6x + 12y = -60 \end{cases}}\ 24y = -72}$$
$$y = -3$$

$\downarrow$

$$6x + 12(-3) = -12 \quad (2)$$
$$6x = 24$$
$$x = 4$$

---

2. By what numbers should you multiply these equations in order to create a zero pair? You do not have to solve the systems.

| | |
|---|---|
| **a.** (1) $\begin{cases} 4x - 7y = 5 \\ 5x + 3y = -11 \end{cases}$<br>(2) | **b.** (1) $\begin{cases} 2x + 3y = 20 \\ 9x + 4y = -1 \end{cases}$<br>(2) |

3. Solve each system of equations. In (a), first multiply the equations as indicated. In (b), you figure out which equation to multiply by what. Then add the equations. Lastly, check your solutions.

| | |
|---|---|
| **a.** $\begin{cases} 2x + 7y = -3 \\ 3x - 2y = 3 \end{cases}$ $\begin{array}{l}\cdot\,3 \\ \cdot\,(-2)\end{array}$<br><br>$\downarrow$<br><br>$+ \begin{cases} \underline{\phantom{x}}x + \underline{\phantom{x}}y = \\ \underline{\phantom{x}}x + \underline{\phantom{x}}y = \end{cases}$ | **b.** $\begin{cases} 6x - 2y = -38 \\ -10x + 5y = 70 \end{cases}$ |

4. Solve each system of equations. Lastly (always!), check your solutions.

a.
$$\begin{cases} 6x - 2y = -20 \\ 12x + 6y = 30 \end{cases}$$

b.
$$\begin{cases} 10x - 6y = 48 \\ (5/3)x - y = 8 \end{cases}$$

c.
$$\begin{cases} 2x + 3y = -19 \\ -3x + 5y = 95 \end{cases}$$

d.
$$\begin{cases} 2x - 5y = -11 \\ 7x + 3y = -100 \end{cases}$$

5. Find the error in the solution of this system of equations. Then correct the error and solve the system.

$$\begin{cases} -7x - 9y = -5 \\ 6x + 8y = 4 \end{cases} \quad \begin{array}{c} \cdot\, 8 \\ \cdot\, 9 \end{array} \quad \rightarrow \quad + \begin{cases} -56x - 70y = -40 \\ 54x + 72y = 36 \end{cases}$$
$$x + 2y = -4$$

6. Find the error in the solution of this system of equations. Then correct the error and solve the system.

(1) $\begin{cases} 10x + 3y = 5 \\ 2x + 7y = 3 \end{cases} \quad \cdot\, 5$
(2)

$\downarrow$

$$\begin{cases} 10x + 3y = 5 \\ 10x + 35y = 15 \end{cases}$$
$$38y = 20$$
$$y = 10/19$$

$\downarrow$

(1) $\quad 10x + 3(\mathbf{10/19}) = 5$
$$10x + 30/19 = 5$$
$$10x = 5 - 30/19$$
$$10x = 65/19$$
$$x = 65/190 = 13/38$$

But the solution (13/38, 10/19) does not check:

$$2(13/38) + 7(10/19) \overset{?}{=} 5$$
$$26/38 + 70/19 \overset{?}{=} 5$$
$$26/38 + 140/38 \overset{?}{=} 5$$
$$166/38 \neq 5$$

7. Kayla DIVIDED instead of multiplying... is that OK?
   Why or why not?

$$\begin{cases} -x + 5y = 11 \\ 3x - 12y = 18 \end{cases} \quad \div\, 3$$

$\downarrow$

   Finish solving the system.

$$+ \begin{cases} -x + 5y = 11 & (1) \\ x - 4y = 6 & (2) \end{cases}$$
$$y = 17$$

8. Solve each system of equations and give the solutions rounded to two decimal digits. Note: in the intermediate steps, use at least five decimal digits. Round *only* the final answers to two decimals.

a. $\begin{cases} 3x + 8y = 10 \\ 15x + 20y = 4 \end{cases}$

b. $\begin{cases} -30x + 40y = 50 \\ 9x - 12y = -15 \end{cases}$

c. $\begin{cases} 0.4x - 0.5y = 0 \\ 2.5x + 0.3y = 1 \end{cases}$

d. $\begin{cases} 40x - 20y = 140 \\ -30x + 15y = 90 \end{cases}$

e. $\begin{cases} -0.3x + 0.7y = -0.5 \\ x + 0.9y = 0 \end{cases}$

f. $\begin{cases} 150x - 0.25y = 17.5 \\ 600x + y = 70 \end{cases}$

Find the values of $a$ and $b$ if the system below has $(1, -3)$ as its solution.

$\begin{cases} 6x - by = 2 \\ ax + by = 3 \end{cases}$

Puzzle Corner

# More Practice

**Example 1.** In the system on the right, we first need to combine like terms, and then rearrange the terms so that both the $x$ and $y$-terms are on the left side.

Only after that are we ready to actually solve the system.

$$\begin{cases} x - 6y = 2(x - 5) \\ 3y = x + 4 \end{cases}$$

$$\downarrow$$

$$\begin{cases} x - 6y = 2x - 10 \\ -x + 3y = 4 \end{cases}$$

$$\downarrow$$

In this case, to continue with the solution from the last step shown here, one could multiply the top *or* the bottom equation by $-1$, to arrive at a zero pair with the $x$-terms. Alternatively, one could multiply the bottom equation by 2, to get the zero pair $-6y$ and $6y$.

$$\begin{cases} -x - 6y = -10 \\ -x + 3y = 4 \end{cases}$$

*You can use a calculator for all the problems in this lesson.*

1. Finish solving the system in example 1.

2. Solve. Use extra paper if necessary. Check your solutions.

**a.** $\begin{cases} 5y = -x + 7 \\ 4(y - 1) = -3(x - 2) \end{cases}$

**b.** $\begin{cases} 3y - 21 = 6(x + 1) \\ x + y + 3 = -4x + 5 \end{cases}$

153

3. Solve.

a. $\begin{cases} 2y + 8x = 51 \\ x + 10y = 2(y + 10x) \end{cases}$

b. $\begin{cases} 5y = 15x - 20 \\ -y - 12 + 13x = 2(5x - 4) \end{cases}$

c. $\begin{cases} x + y + 5 = -3x + 46 \\ -3(y + 7) + 2x = 3(x + 7) \end{cases}$

d. $\begin{cases} 3(x + y) + 50 = 4x - 65 - 2y \\ 3(x - 15) + 5y = 4y - 20 \end{cases}$

4. Find a way of multiplying the **2nd equation** in each system that makes the solution easy.

a. $\begin{cases} 3x + y = 6 \\ -5x - \dfrac{y}{2} = 4 \end{cases}$

b. $\begin{cases} -x + 9y = 7 \\ \dfrac{x}{3} - 2y = 1 \end{cases}$

5. Solve. Start out by multiplying the equations to eliminate the fractions.

a. $\begin{cases} 2x + \dfrac{y}{2} = 1 \\ x - \dfrac{y}{3} = 4 \end{cases}$

b. $\begin{cases} \dfrac{x}{3} + y = 3 \\ \dfrac{2x}{3} - 2y = -1 \end{cases}$

6. Use these equations for more practice.

a. $\begin{cases} 10 = \dfrac{1}{4}(x+y) \\[2mm] x - \dfrac{y}{3} = 4 \end{cases}$

b. $\begin{cases} 6x - y = \dfrac{4}{5} \\[2mm] \dfrac{y}{2} - \dfrac{3x}{4} = 1 - y \end{cases}$

7. Here are more practice problems. Use a notebook. In (d), use decimals.
   *Check also the worksheet generator for more practice problems.*

a. $\begin{cases} x + y = 2 \\[1mm] 5y - 7 = 10x - 6 \end{cases}$

b. $\begin{cases} -45x + 10y = -39 \\[1mm] 15x + 30y = -27 \end{cases}$

c. $\begin{cases} -1 = 2x - y \\[1mm] 5y - 15 = 2(5x + 1) \end{cases}$

d. $\begin{cases} \dfrac{1}{2}x + y = 70 \\[2mm] -5x + y = -30 \end{cases}$

e. $\begin{cases} 4(y + 2) = 3(4x - 1) \\[2mm] y + 3x + 2 = x + \dfrac{1}{2} \end{cases}$

f. $\begin{cases} \dfrac{1}{3}(x + 5) + y = 30 \\[2mm] -y - 30 = 2(3x - y + 15) \end{cases}$

g. $\begin{cases} x + 2y - 16 = -1 \\[2mm] y - 6 = \dfrac{1}{4}(x + 9) \end{cases}$

h. $\begin{cases} y + 3x = -\dfrac{3}{2} \\[2mm] 2x + y = 12x + 5 \end{cases}$

# Applications, Part 2

**Example 1.** The digit sum of a two-digit number is 8, and the value of the number is seven times its ones digit. Find the number.

Let $t$ be the tens digit, and $u$ be the ones (units) digit. (We're not using $o$ as a variable as that is so easy to confuse with the number 0.) Note that both $t$ and $u$ are single-digit numbers ($t \neq 0$).

The digit sum is 8, which means $t + u = 8$.

The other clue we have is trickier to translate into an equation. Keep in mind that $t$ is just a digit — a number from 1 through 9. The actual *value* of our two-digit number is given by $10t + u$.

So, the second clue is that $10t + u = 7u$.

Now that we have the equations, it is quite easy to solve the system formed by them.

1. Solve the system of equations from example 1.

2. The digit sum of a two-digit number is 11, and the tens digit is three times the ones digit minus 1. Find the number.

3. The digit sum of a two-digit number is 9. If you reverse the digits, the number formed is nine more than the original number. Find the number.

4. The tens digit of a two-digit number is four more than its ones digit. The value of the number is seven times its digit sum. Find the number.

5. The ones digit of this two-digit number is three more than its tens digit. If you reverse the digits, the number formed is two more than twice the original number. Find the number.

6. Make your own two-digit number puzzle with two clues for someone to solve.

7. A store sells two kinds of energy bars: fruit bars for $8 each and nut bars for $10 each. Let $f$ be the number of fruit bars and $n$ the number of nut bars. These two equations have to do with the sales on one particular day:

$$8f + 10n = 264$$

$$n = f + 3$$

**a.** What does the first equation signify in practical terms?

**b.** And the second?

**c.** Solve the system of equations. What will you find out?

8. Emma and Daisy each bought a flower bouquet from a local flower shop. Emma's bouquet had 12 roses and 10 lilies, and cost $39. Daisy's bouquet had 8 roses and 16 lilies and cost $40. How much does one rose cost? One lily?

9. The perimeter of a rectangle is 472 m, and one side is 56 m longer than the other. Find the side lengths of the rectangle.

10. A solar supply store sells two different kinds of solar panels, at two different prices. If you buy 20 of the cheaper kind and 10 of the other, the total cost is $3300. If, instead, you buy 10 of the cheaper kind and 20 of the other, the total cost is $3900. Find the price of each kind of panel.

The area of a rectangle is 3364 square units, and one of its sides is four times the other. Find the perimeter of the rectangle.

**Puzzle Corner**

159

# Speed, Time, and Distance Problems

There are many jokes about algebra word problems where a train leaves a station at a certain hour. You can now solve these types of problems with your knowledge of systems of equations. One of the most effective ways to do so is to first build a chart.

**Example 1.** A train leaves a station at 9:00 AM and travels with a constant speed of 90 km/h. Another train leaves the same station 10 minutes later, travelling to the same direction at the speed of 100 km/h. At what time will the second train reach the first?

We will be using the formula $d = vt$ extensively in these problems. Let's build a chart. The goal is to have TWO, not three or more, variables present in the chart. The formula $d = vt$ has three variables, and since the speed, distance, and time can be different for each train, theoretically we could have six variables. However, invariably, the problem gives information for one or some of these variables, and something about the situation means that the distance or the time or the speed is the same for both trains.

To get started, we gather some information in the chart. The distance that train 1 and train 2 travel until they meet is the same, so that is why we use the same variable, $d$, for it.

|         | *distance* | *velocity* | *time* |
|---------|------------|------------|--------|
| Train 1 | $d$        | 90 km/h    | $t_1$  |
| Train 2 | $d$        | 100 km/h   | $t_2$  |

The times ($t_1$ and $t_2$) are different, but we do know that they differ by 10 minutes, so, actually we will get by using only *one* variable for time, like this:

|         | *distance* | *velocity* | *time*   |
|---------|------------|------------|----------|
| Train 1 | $d$        | 90 km/h    | $t$      |
| Train 2 | $d$        | 100 km/h   | $t - 10$ |

The chart now contains only two variables. However, we have one more thing to change. The speed is in km/h, whereas the 10 has to do with minutes. For our equation to work, the time units need to be the same, so we will change the 10 to 1/6 (in hours).

|         | *distance* | *velocity* | *time*    |
|---------|------------|------------|-----------|
| Train 1 | $d$        | 90 km/h    | $t$       |
| Train 2 | $d$        | 100 km/h   | $t - 1/6$ |

The equations always follow the same formula: $d = vt$, and we use that same formula for both Train 1 and Train 2. So, the two equations we get are:

$$\begin{cases} d = 90t \\ d = 100(t - 1/6) \end{cases}$$

The quickest way to solve this system is to set $90t$ equal to $100(t - 1/6)$ and solve for $t$.

1. Solve the system of equations from example 1 and answer the question: At what time will the second train reach the first? Is the answer surprising?

$$\begin{cases} d = 90t \\ d = 100(t - 1/6) \end{cases}$$

2. Your friend starts walking at a speed of 6 km/h from your home to his. Exactly 15 minutes later, you decide you want to join him so you take your bicycle and start after him, with a speed of 18 km/h. How far have you ridden by the time you reach your friend?

|  | distance | velocity | time |
|---|---|---|---|
| Your friend |  |  |  |
| You |  |  |  |

3. A tortoise and hare race a distance of 100 m. The hare gives the tortoise a 10-minute lead time. Then he quickly runs the 100 metres and wins the race. After the hare has finished, the tortoise takes an *additional* 6 minutes to reach the finish line. If the speed of the hare is 15 m/s, find the time the tortoise takes to finish the race and the tortoise's speed.

*Hint: since the speed is in metres per second, and the distance is in meters, the time unit will be <u>seconds</u>.*

|  | distance | velocity | time |
|---|---|---|---|
| Tortoise |  |  |  |
| Hare |  |  |  |

**Example 2.** A train leaves Turin (Italy), heading for Milan (Italy), a distance of 125 km, at 1 PM and travels with a constant speed of 90 km/h. Another train leaves Milan, heading for Turin and travelling at a constant speed, at the same time. They meet 45 minutes later. (We hope they don't crash!) What is the speed of the second train? What distance has the second train travelled by that time?

We fill our chart again. The time, 45 minutes, is 3/4 hour. The speed of the second train, $v_2$, is unknown.

There is a relationship between the two distances, because $d_1 + d_2 = 125$ km. So, we can get by with just one variable for the distance:

|         | distance  | velocity  | time |
|---------|-----------|-----------|------|
| Train 1 | $d$       | 90 km/h   | 3/4  |
| Train 2 | $125 - d$ | $v_2$     | 3/4  |

Our equations are:
$$\begin{cases} d = 90(3/4) \\ 125 - d = (3/4)v_2 \end{cases}$$

Here, it is handy to use the substitution method, since we have an expression for $d$. So, we substitute $90(3/4)$, which equals 67.5, in place of $d$ in the second equation:

$$\begin{aligned}
(2) \quad 125 - \mathbf{67.5} &= (3/4)v_2 \\
57.5 &= (3/4)v_2 \qquad |\cdot 4 \\
230 &= 3v_2 \\
v_2 &= 76.\overline{6}
\end{aligned}$$

So, the speed of the second train is $76.\overline{6}$ km/h. The problem also asked what distance the second train has travelled by that time. To find that, we use the formula $d = vt$:  $d = 76.\overline{6}$ km/h $\cdot$ (3/4 h) = 57.5 km.

4. Two trains leave the same station at the same time, one travelling due east and the other travelling due west. Train 1 travels at a speed of 120 km/h and Train 2 at the speed of 100 km/h. When are the trains 50 km apart from each other?

|         | distance | velocity | time |
|---------|----------|----------|------|
| Train 1 |          |          |      |
| Train 2 |          |          |      |

5. Train 1 leaves the station heading due south and Train 2 leaves the same station at the same time, heading due north. Train 1 travels at the speed of 110 km/h. After 30 minutes, the trains are 120 km apart. How fast is Train 2 travelling?

|         | distance | velocity | time |
|---------|----------|----------|------|
| Train 1 |          |          |      |
| Train 2 |          |          |      |

6. **a.** Two horses, Ranger and Chip, start racing at the same time. Ranger runs at a steady speed of 16 m/s. After 100 seconds, they are 600 m apart from each other, Ranger leading. How fast is Chip running?

|        | distance | velocity | time |
|--------|----------|----------|------|
| Ranger |          |          |      |
| Chip   |          |          |      |

**b.** What would Chip's speed need to be, so that after 100 seconds, he would only be 50 m behind Ranger?

**Example 3.** A motorboat travels downstream on the river a distance of 8 km, in 20 minutes. Doing the same trip upstream takes it 4 minutes longer. How fast is the river flowing? What is the speed of the boat in still water?

Our chart method will still work. We are dealing with two speeds: that of the boat ($v_b$) (in still water), and that of the water ($v_w$) in the river. Going downstream, the boat's actual speed is its own speed PLUS the speed of the water. Going upstream, it has to fight the current and its actual speed is $v_b - v_w$.

This is what the chart looks like. Using these quantities, the speeds will end up being in km per minute.

|            | distance | velocity      | time   |
|------------|----------|---------------|--------|
| Downstream | 8 km     | $v_b + v_w$   | 20 min |
| Upstream   | 8 km     | $v_b - v_w$   | 24 min |

Our equations are:
$$\begin{cases} 8 = 20(v_b + v_w) \\ 8 = 24(v_b - v_w) \end{cases}$$

7. Solve the problem in example 3.

8. An airplane travels from City A to City B in 2 hours 30 minutes, flying with the wind, and in 2 hours 45 minutes flying against the wind. If the speed of the airplane in still air is 900 km/h, find the distance between the cities to the nearest 10 km.

9. With a tailwind, an airplane can fly from City 1 to City 2, a distance of 650 km, in 40 minutes. If the speed of the wind is 30 km/h, find the time the plane takes to fly the same distance against the wind.

10. You swim downstream, from a dock to a certain rock in the middle of the river, in 43 seconds. Swimming back (upstream) takes you 10 seconds longer. If your swimming speed in still water is 1.6 km/h, what is the speed of the water in the river?

**Puzzle Corner**      The following are "trick" problems. Have fun!

**(1)** Train 1 leaves Jackson at 1:30 PM, travelling at 95 km/h towards Atlanta, a distance of 560 km, and Train 2 leaves Atlanta, heading towards Jackson at the same time, travelling at 105 km/h. When they meet, which train is closer to Atlanta?

**(2)** At 6:30 PM, you board a train in Dallas, heading south, and 10 minutes later, your friend boards a train at Kansas City, heading north. If both trains travel at 100 km/h, when do you pass each other?

**(3)** Two trains leave a station at the same time, one heading east, the other heading west. After 15 minutes, they are 64 km apart. Which train is travelling faster?

# Mixtures and Comparisons

The chart method lends itself well also for solving problems about mixtures.

**Example 1.** Raisins cost $8/kg and almonds cost $13.70/kg. Ashley made a mixture of both so that the cost of 1 kilogram is $10.00. What amount of the mixture is raisins?

Let $r$ be the weight of the raisins and $a$ be the weight of the almonds in the mixture (in kilograms). We can fill in a chart:

|         | weight (kg) | cost per kg | cost ($) |
|---------|-------------|-------------|----------|
| Raisins | $r$         | $8/kg       | $8r$     |
| Almonds | $a$         | $13.70/kg   | $13.7a$  |
| Mixture | 1           | $10/kg      | $10.00   |

There are *two* unknowns, $r$ and $a$. Now all we need is *two* equations. What allows us to build those equations? What is equal to what? Think about it first. Then check the end of the lesson for the two equations. The rest of the solution is left for an exercise.

1. Finish solving the problem in Example 1.

2. A mixture of grains meant for a bird feed is 16% protein by weight. It consists of corn (12% of protein by weight) and chickpeas (22% protein by weight). Find out how much corn and how much chickpeas is in 2 kg of this mixture.

   The given information is already put in the chart below. Choose two variables and fill in the chart. Then write two equations using the information in the chart. Check your equations before going to exercise #3.

|           | weight (kg) | protein (%) | protein (weight) |
|-----------|-------------|-------------|------------------|
| Corn      |             | 12          |                  |
| Chickpeas |             | 22          |                  |
| Mixture   | 2           | 16          | 0.16(2)          |

   Equations:

3. Solve the system of equations from #2.

4. You're an owner of a bakery, and one thing your bakery produces is honey buns. The recipe calls for 2 kg of honey. You're planning to use both expensive wildflower honey and cheaper regular honey, so you can mention on the packaging that the buns contain wildflower honey. But, you don't want to use very much of the expensive honey to keep the price reasonable. The wildflower honey costs $50/kg and the regular honey $20/kg.

How much of each kind of honey should you use in the recipe so that the total cost of honey will be $25/kg (in other words, $50 for the entire recipe)?

5. Brass is an alloy of copper and zinc. There exist many different kinds of brasses where copper and zinc are in different proportions. If you have the two alloys listed below on hand, how much of each would you need to melt together in order to produce 1000 kg of brass that contains 80% copper and 20% zinc?

| | copper | zinc |
|---|---|---|
| Alloy 1 | 90% | 10% |
| Alloy 2 | 65% | 35% |

6. Hydrogen peroxide ($H_2O_2$) comes in different concentrations, containing hydrogen peroxide and water. You have the 10% and 3% solutions of $H_2O_2$ on hand, but you want to have an 8% solution of $H_2O_2$. How much of each solution do you need in order to get one liter of 8% solution of $H_2O_2$?

7. John is comparing two different "Subscribe and save" plans for energy bars his family consumes regularly.

| Subscribe & save! PLAN 1 | Subscribe & save! PLAN 2 |
|---|---|
| • An annual fee of $20<br>• Each energy bar costs $3.60 | • An annual fee of $50<br>• Each energy bar costs $3 |

**a.** Write a linear equation to represent the cost (C) of $x$ energy bars under each plan.

Plan 1:  C =                                          Plan 2:  C =

**b.** Now consider the two equations above as forming a system of equations, with C and $x$ being the two variables, and solve the system.

**c.** What does the solution signify?

**d.** Graph the equations.

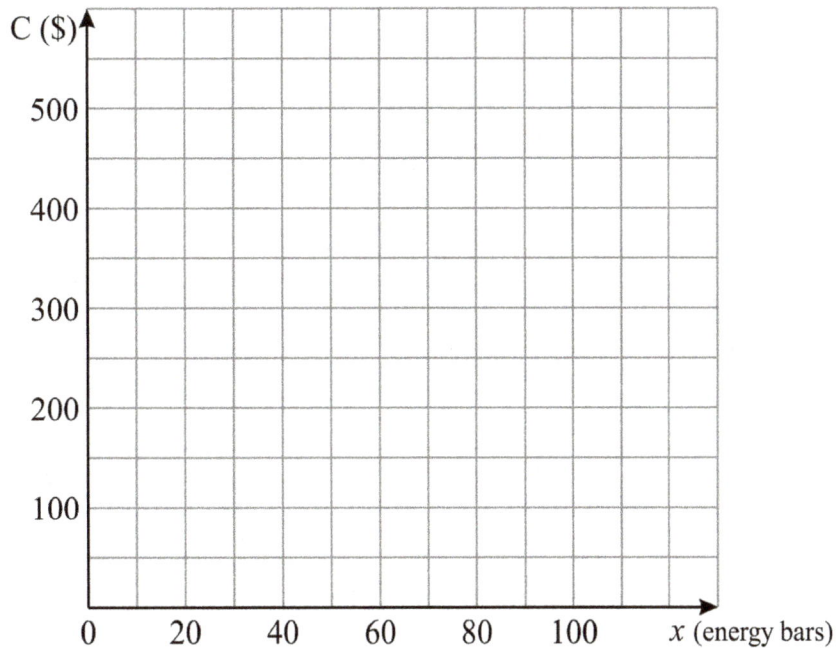

**e.** For what amounts of energy bars will plan 1 will be more cost-efficient?

For what amounts will plan 2 be so?

8. A gym offers two membership plans. Plan 1 is simply a \$45 monthly fee. Plan 2 has an initial \$100 membership fee, and a \$20 monthly fee.

**a.** Write a linear equation to represent the cost (C) of $t$ months of gym membership under each plan.

Plan 1:  C =                                   Plan 2:  C =

**b.** Now consider the two equations above as forming a system of equations, with C and $t$ being the two variables, and solve the system.

**c.** What does the solution signify?

**d.** Graph the equations. Design a scaling for the vertical axis so that the graphs will fit.

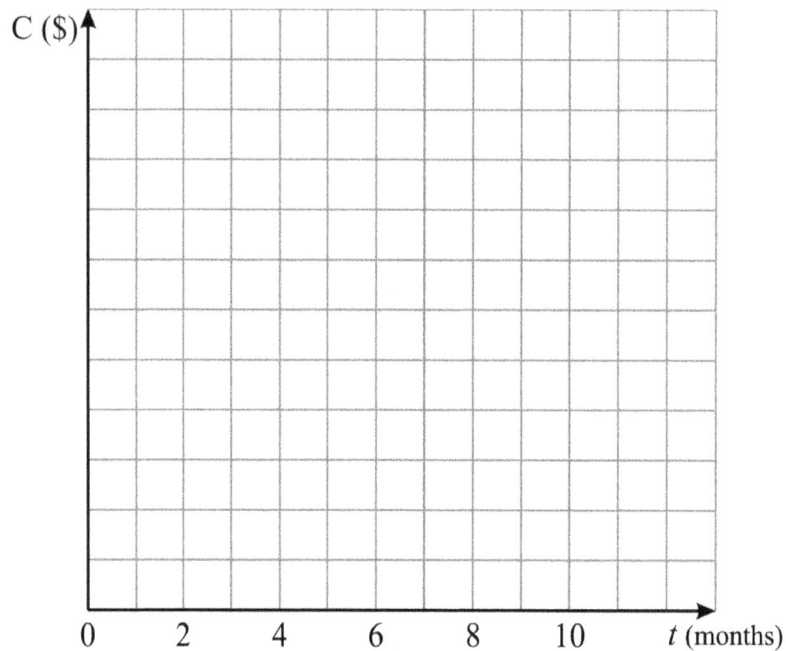

**e.** In what situation will Plan 1 be more cost-efficient than Plan 2?

---

# Mixed Review Chapter 7

1. Show that the two trapezoids are similar by describing a sequence of transformations that could map trapezoid ABCD to the smaller trapezoid.

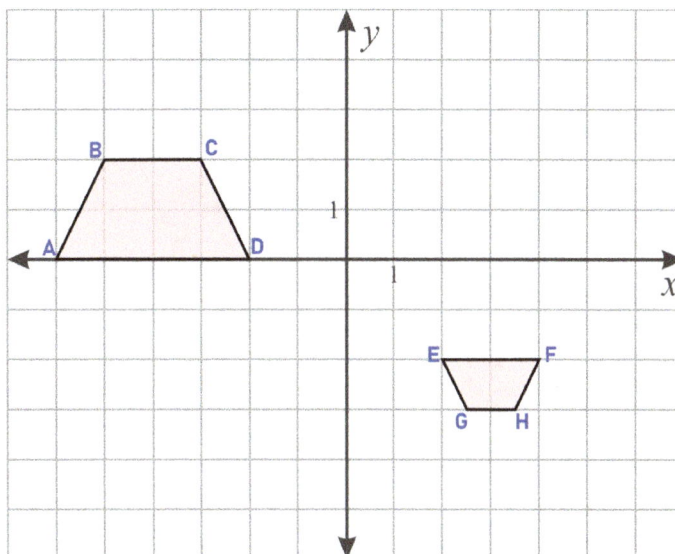

2. Solve the equations.

| | |
|---|---|
| **a.**  $s + 2 - 5s \ = \ 3 - 8(s - 1) - 6s$ | **b.**  $6 + \dfrac{2x - 5}{3} \ = \ x - 2$ |
| **c.**  $0 \ = \ \dfrac{x - 5}{3} + \dfrac{x + 5}{4}$ | **d.**  $10 \ = \ 2y + \dfrac{2 - 3y}{6}$ |

3. Give an example of each type of equation that has $2x - 5$ on the left side of the equation.

**a.** No solutions:

$2x - 5\ =$

**b.** One solution:

$2x - 5\ =$

**c.** An infinite number of solutions:

$2x - 5\ =$

4. Find the equation of each line.

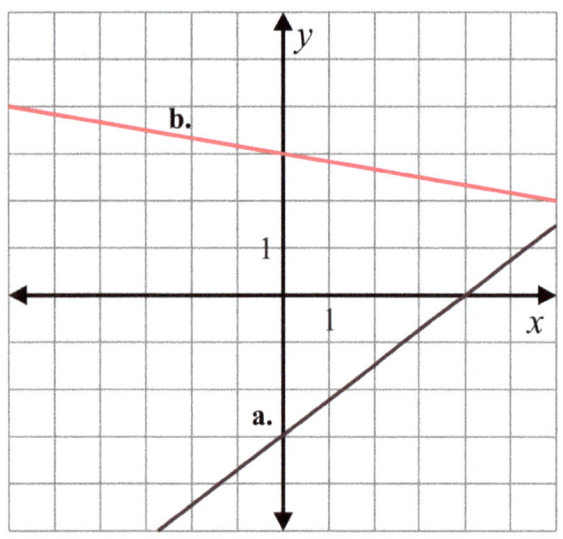

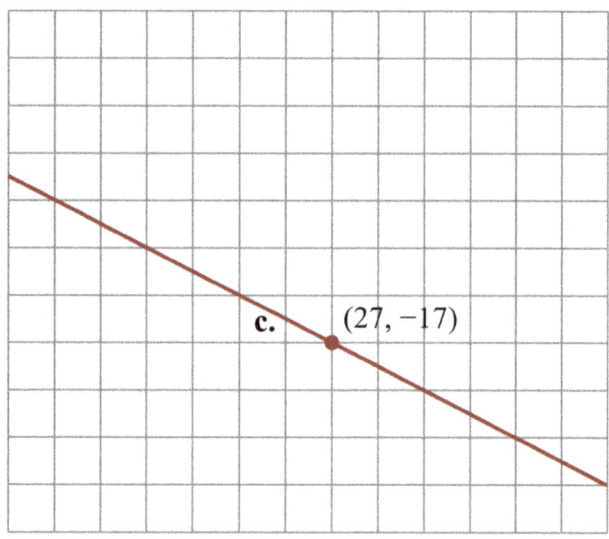

**a.**

**b.**

**c.**

5. Solve each system of equations.

**a.** $\begin{cases} y = -2(x + 10) \\ 2(y - x) = 8 \end{cases}$

**b.** $\begin{cases} 2x - 4y = -3 \\ 5x + 3y = 1 \end{cases}$

6. **a.** Two dogs, Rocky and Charlie, start racing at the same time. Rocky runs at a steady speed of 10 m/s. After 20 seconds, they are 8 m apart from each other, Rocky leading. How fast is Charlie running?

| | distance | velocity | time |
|---|---|---|---|
| Rocky | | | |
| Charlie | | | |

**b.** What would Charlie's speed need to be, so that after 20 seconds, he would only be 2 m behind Rocky?

7. Brass is an alloy of copper and zinc. There exist many different kinds of brasses where copper and zinc are in different proportions. If you have the two alloys listed below on hand, how much of each would you need to melt together in order to produce 5,000 kg of brass that contains 75% copper and 25% zinc?

| | copper | zinc |
|---|---|---|
| Alloy 1 | 90% | 10% |
| Alloy 2 | 65% | 35% |

8. An elephant runs (or "fast-walks") at a constant speed of 6 m/s.

   **a.** Convert this speed to metres per minute.

   **b.** Write an equation for the distance ($d$) the elephant covers as a function of time in *minutes* ($t$).

   **c.** Plot your equation. Design the scaling of the vertical axis so that the point for three minutes fits on the graph.

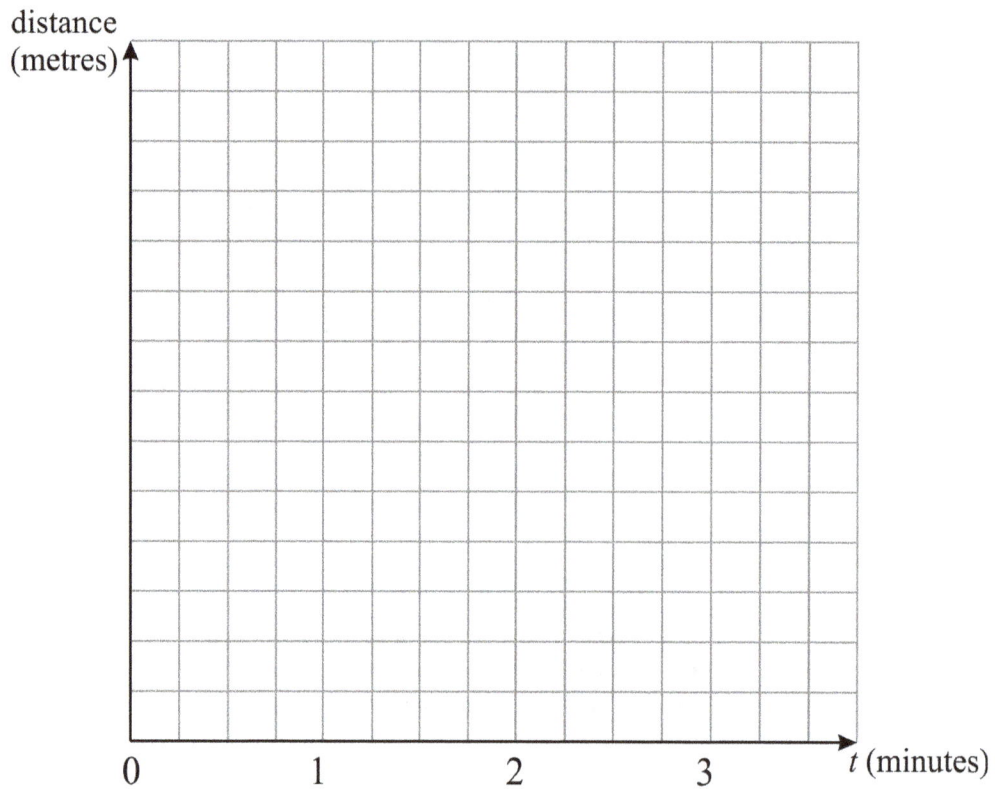

   **d.** What is the slope of the line in your graph?

   **e.** The table below shows the distance and time when Elizabeth bicycles at a constant speed.

| Time (min) | 2 | 4 | 5 | 10 | 15 |
|---|---|---|---|---|---|
| Distance (m) | 800 | 1600 | 2000 | 4000 | 6000 |

   Who goes faster, the elephant or Elizabeth?

   **f.** How much larger distance does the faster one cover in 3 minutes?

# Chapter 7 Review

1. Tell how many solutions each system of equations has by inspecting the equations. You do not have to find the solution(s).

a. $\begin{cases} y + 2x = -8 \\ 2y + 4x = -16 \end{cases}$  b. $\begin{cases} 7x - 2y = -1 \\ -2y + 7x = 3 \end{cases}$  c. $\begin{cases} x + y = -1/2 \\ 2y - 3x = 6 \end{cases}$

2. Solve the systems of equations depicted by the graphs.

**a.**

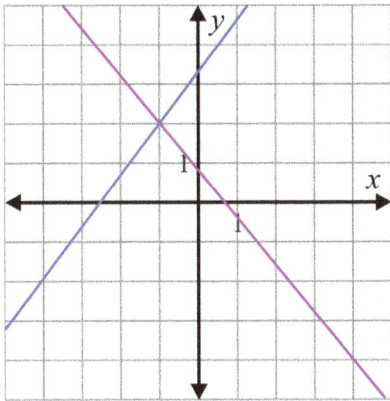

**b.**

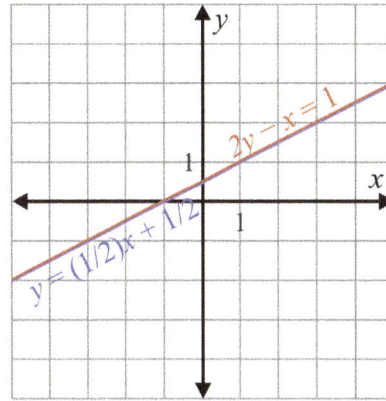

3. Find the system below that has a single solution, and solve it.

a. $\begin{cases} 4x - 2y = 1 \\ y = 2x + 6 \end{cases}$  b. $\begin{cases} y = (1/2)x + 6 \\ x + 6y = 0 \end{cases}$  c. $\begin{cases} 3(x + y) = 3 \\ y = -x + 1 \end{cases}$

4. Solve the system of equations by substitution. Then graph the lines. Verify that the intersection point of the lines is the solution you found algebraically.

$$\begin{cases} x - 2y = 3 \\ 2x + y = -8 \end{cases}$$

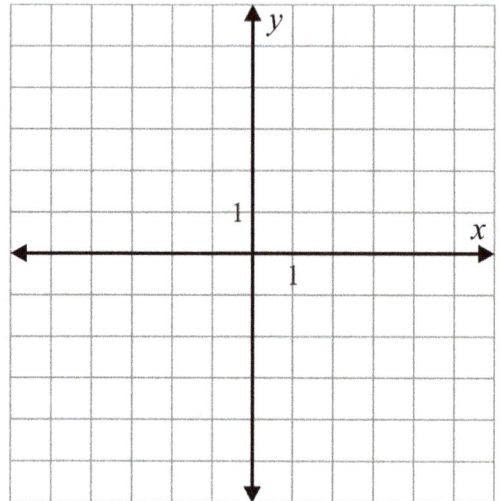

5. Solve each system of equations.

a. $\begin{cases} x = -(y - 5) \\ 2y = 12(1 - x) \end{cases}$

b. $\begin{cases} 4.5x - 3y = 0 \\ 7x + 2y = -2 \end{cases}$

c. $\begin{cases} 3x = -9(y + 1/3) \\ -2x - 6y = 2 \end{cases}$

d. $\begin{cases} 5x - 3(y + 2) = 0 \\ -5x + 6y = 7 \end{cases}$

6. Find the error in the solution of this system of equations. Then correct the error and solve the system.

(1) $\begin{cases} 6x + 3y = 3 \\ -2x + 9y = 11 \end{cases} \cdot 3$
(2)

$\downarrow$

$\begin{cases} 6x + 3y = 3 \\ -6x + 27y = 33 \end{cases}$
$\overline{\phantom{0000000000}}$
$\qquad 30y = 36$
$\qquad\quad y = 5/6$

$\downarrow$

(1) $6x + 3(5/6) = 3$
$\qquad 6x + 5/2 = 3$
$\qquad\qquad 6x = 3 - 5/2 = 1/2$
$\qquad\qquad 6x = 1/2$
$\qquad\qquad\; x = 1/12$

However, the solution (1/12, 5/6) does *not* fulfill the 2nd equation:

$-2(1/12) + 9(5/6) \overset{?}{=} 11$

$-1/6 + 45/6 \overset{?}{=} 11$

$44/6 \neq 11$

7. Solve each system of equations and give the solutions as decimals, rounded to three decimal digits.
   Note: if you use a decimal in the intermediate steps, include at least five decimal digits. Round only the final answers to three decimals.

a. $\begin{cases} 7x + 6y = -1 \\ 11x + 2y = 3 \end{cases}$

b. $\begin{cases} 3.4x + 0.7y = 5 \\ 0.5x - 0.2y = -2 \end{cases}$

8. A bicycle shop sells both bikes and trikes. If there are 80 wheels and 31 vehicles, find the number of trikes the shop has.

9. Denny and Sammy were comparing their ages. Sammy said, "In four years, my age will be 2/3 of your age." Denny said, "And I am 14 years younger than double your age." Find their ages.

10. The digit sum of a two-digit number is 11. The number itself (its value) is two less than seven times its ones digit. Find the number.

11. A mixture of peanut butter and protein powder weighs 210 grams and contains 30% protein. The protein powder by itself contains 90% protein, and the peanut butter by itself contains 25% protein. Find how many grams of peanut butter and how many grams of protein powder are in the mixture.

12. Flying with the wind, a crow flies from his favourite tree to a nearby pond, a distance of 400 metres, in 48 seconds. Coming back, flying against the wind, it takes him 72 seconds. What is the crow's speed in still air? What is the wind speed?

# Chapter 8: Bivariate Data
## Introduction

The last chapter of grade 8 covers statistical topics that have to do with bivariate data, or data involving two variables.

The first lesson introduces scatter plots. Students analyse the data in a variety of scatter plots, and determine visually whether there is an association between the variables. Next, they learn about basic patterns we often see in scatter plots, such as positive and negative association, linear association, clusters, and outliers. They also make scatter plots from given data, describe any special features in the plot, and answer a variety of questions related to the data.

In the following lesson, students fit a line (informally) to the data points displayed in a scatter plot. Mathematicians have developed several algorithms for finding a line of best fit, such as linear regression, but we are not using those here. Students use the basic idea of trying to leave close to an equal number of points on each side of the line, and also judging the fit by the closeness of the points to the line. This resembles the thought behind the linear regression algorithm, which finds the line of best fit by minimising the squares of the distances of the data points to the line.

The last topic relating to scatter plots is the equation of the trend line. Students use the equation of the trend line to solve problems in the context of the data, interpreting the slope and intercept of the equation.

Then we turn our attention to categorical bivariate data, that is, data involving two variables that may or may not be numerical, but is divided into categories. Students learn that bivariate categorical data can be summarised in a two-way table, and if there is a pattern of association between the variables, it can be seen in the table.

Students construct and interpret two-way tables summarising data on two categorical variables. In the last lesson, they calculate relative frequencies for rows or columns, and use those to describe the possible association between the two variables.


## Pacing Suggestion for Chapter 8

This table does not include the chapter test as it is found in a different book (or file).
Please add one day to the pacing if you use the test.

| The Lessons in Chapter 8 | page | span | suggested pacing | your pacing |
|---|---|---|---|---|
| Scatter Plots ............................................................ | 183 | *3 pages* | 1 day | |
| Scatter Plot Features and Patterns ........................................ | 186 | *4 pages* | 1 day | |
| Fitting a Line ........................................................ | 190 | *4 pages* | 1 day | |
| Equation of the Trend Line ............................................... | 194 | *5 pages* | 1-2 days | |
| Two-Way Tables ........................................................ | 199 | *3 pages* | 1 day | |
| Relative Frequencies ................................................... | 202 | *5 pages* | 2 days | |
| Mixed Review Chapter 8 ................................................. | 207 | *5 pages* | 2 days | |
| Chapter 8 Review ...................................................... | 212 | *3 pages* | 1 day | |
| Chapter 8 Test (optional) | | | | |
| **TOTALS** | | *32 pages* | 10-11 days | |

## Helpful Resources on the Internet

We have compiled a list of Internet resources that match the topics in this chapter, including pages that offer:

- **online practice** for concepts;
- online **games**, or occasionally, printable games;
- **animations** and interactive **illustrations** of math concepts;
- **articles** that teach a math concept.

We heartily recommend you take a look! Many of our customers love using these resources to supplement the bookwork. You can use these resources as you see fit for extra practice, to illustrate a concept better and even just for some fun.  Enjoy!

https://l.mathmammoth.com/gr8ch8

Scan me

# Scatter Plots

A **scatter plot** depicts **bivariate data**, meaning that the data involves **two variables**. In the scatter plot below, the variables are the husband's age and the wife's age. Each dot in this scatter plot represents a husband-wife couple. In other words, the coordinates of the dot give us the ages of the husband and the wife.

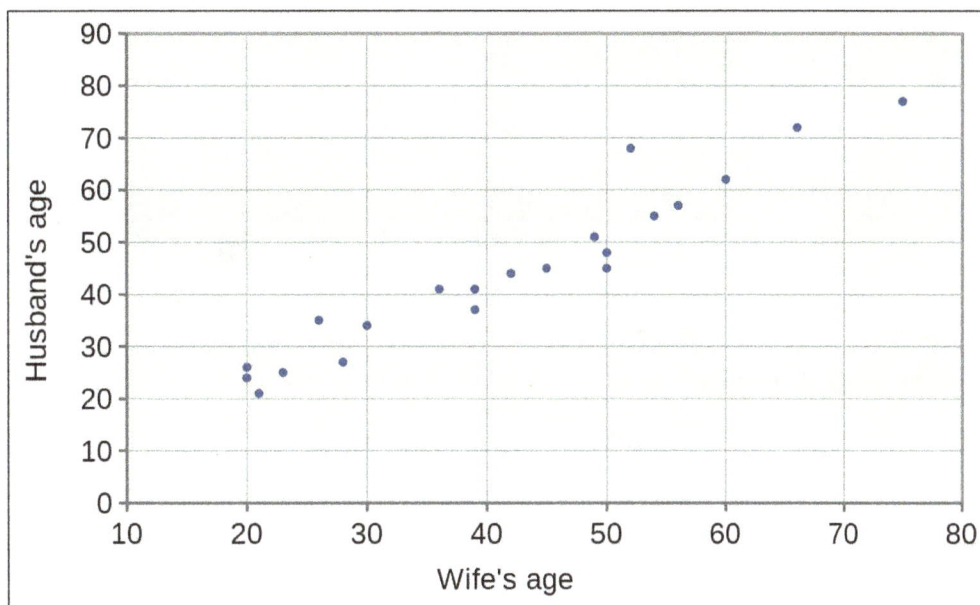

1. Refer to the scatter plot above.

   **a.** Locate the dot with coordinates (36, 41). What does it signify?

   **b.** Find two couples where the wife is the same age in both cases. Estimate the ages of their husbands.

   **c.** Find the couple with the third oldest husband in this data set. How old is his wife?

   **d.** Is it true that the youngest wife is married to the youngest husband? Explain.

   **e.** Is it true that the oldest wife is married to the oldest husband? Explain.

   **f.** Do you notice a relationship between the two variables? Explain what you see.

2. The scatter plot below shows the final grade and the number of missed classes for a group of students who took a 15-week course on nursing care.

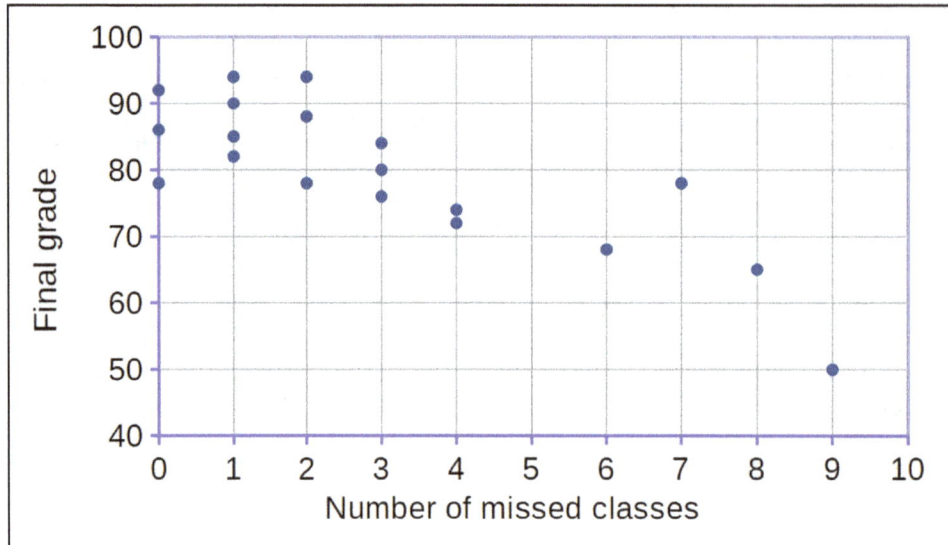

a. What are the two variables depicted?

b. How many times did the two students with the highest grade miss class?

c. Add to the plot a point depicting a student with nearly perfect attendance but with a mediocre grade (from 60 to 75).

d. Do you notice a relationship between the two variables? Explain what you see.

e. Is the relationship *causal*? In other words, is the number of missed classes directly causing the student's final grade?

3. Each dot in this plot depicts a bag of rice.

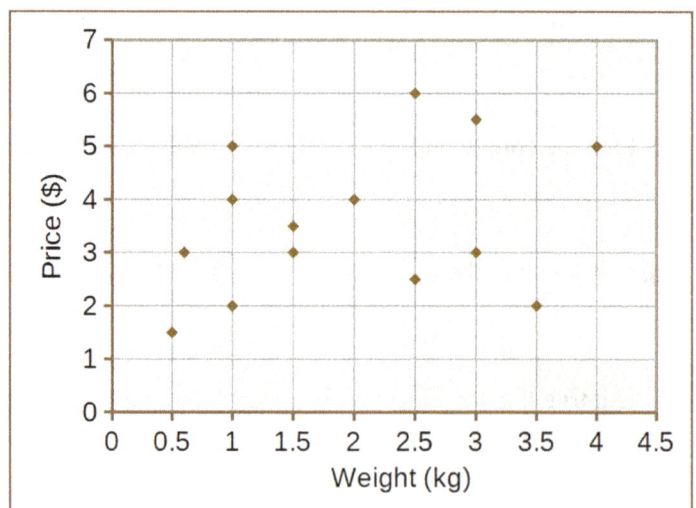

a. What does the point (2.5, 6) signify?

b. Find two bags of rice that cost $2 per kilogram.

c. Add a point for a 4-kg bag of rice with a very low price.

d. Which bag gives you the best value for your money?

e. Could we say that as the weight increases, the price also increases?

4. The data below shows the shots and the goals for a mini-tournament of four games for each player on a soccer team. (A shot is when the player kicks the ball towards the goal.) Each row has the data for one player in a team.

a. Make a scatter plot of the data. Choose the scaling for both axes wisely, so that all your data fits in the graph. Note that the scaling does not have to start from 0. See question 2 for an example where the vertical axis started at 40.

| Shots | Goals |
|-------|-------|
| 20    | 2     |
| 21    | 1     |
| 21    | 2     |
| 22    | 3     |
| 22    | 2     |

| Shots | Goals |
|-------|-------|
| 22    | 3     |
| 24    | 4     |
| 24    | 2     |
| 25    | 4     |
| 25    | 5     |

| Shots | Goals |
|-------|-------|
| 26    | 3     |
| 26    | 2     |
| 27    | 4     |
| 28    | 6     |
| 30    | 5     |

## Shots vs. Goals for Four Games

Goals

Shots

b. Do you notice a relationship between the two variables? Explain what you see.

c. Which players do you feel are the best? Explain why.

185

# Scatter Plot Features and Patterns

When studying data in two variables, we are usually interested in knowing whether there is any association between them; in other words, whether the two things we are studying are connected in any way.

An association will be seen visually in a scatter plot when the points lie in a visible pattern. The data can also have outliers and clusters.

### A positive linear association

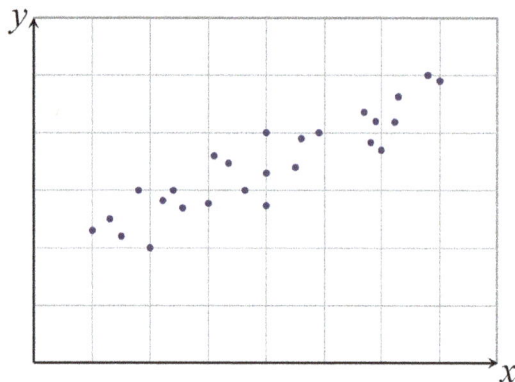

The pattern of the points is as if on a line (linear). As the $x$ values increase, the $y$-values increase also, so the association is **positive**.

### A negative linear association

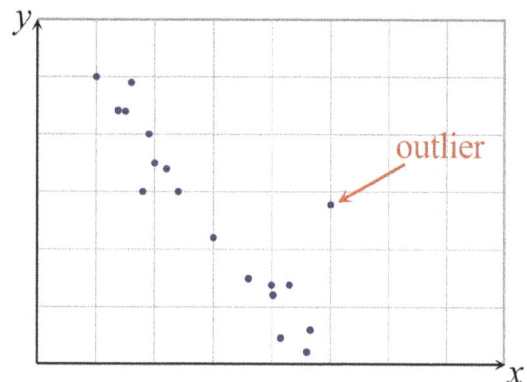

The points lie in a linear pattern. As the $x$ values increase, the $y$-values *decrease*, so the association is **negative**. One point lies far from the others; it is an **outlier**.

### A positive nonlinear association

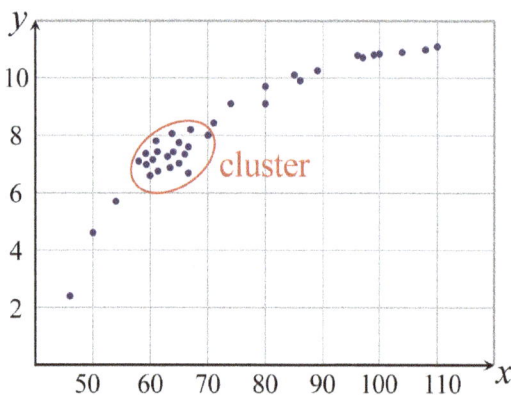

Here as the $x$-values increase, the $y$-values increase also, so the association is positive. But the pattern of dots follows a curve, not a straight line, so we say that the association is **nonlinear**.

Additionally, we see a **cluster** — many points close together in a small area, around the $x$ values of 58 to 70 and $y$-values of 7.5 to 8.3.

### No association

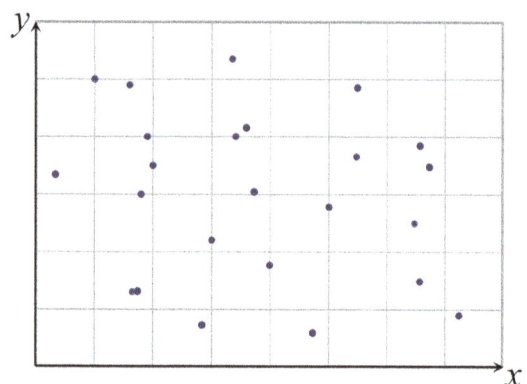

Here, there is no association between the variables. The dots are scattered and do not lie in any visible pattern.

1. Describe the patterns and features you see in each scatter plot.

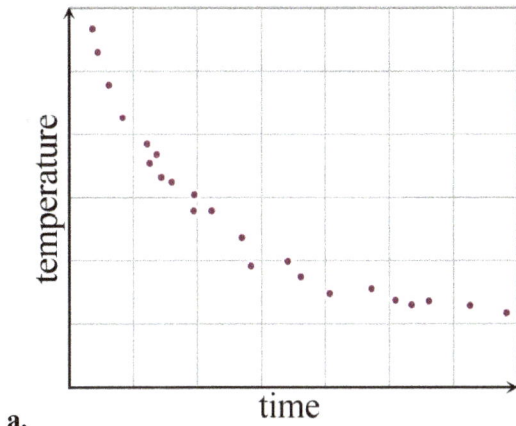

a. (temperature vs. time)

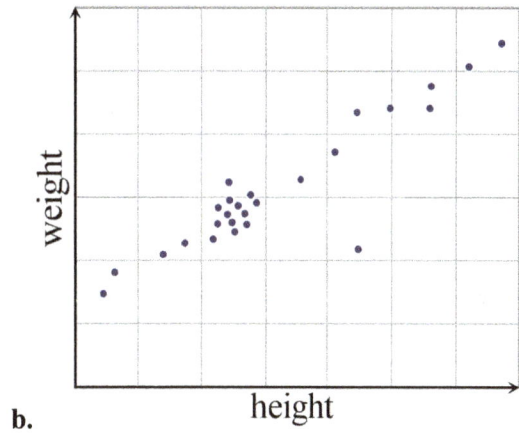

b. (weight vs. height)

2. This graph shows the mass and fuel consumption (in litres per 100 km, for highway driving) of various cars.

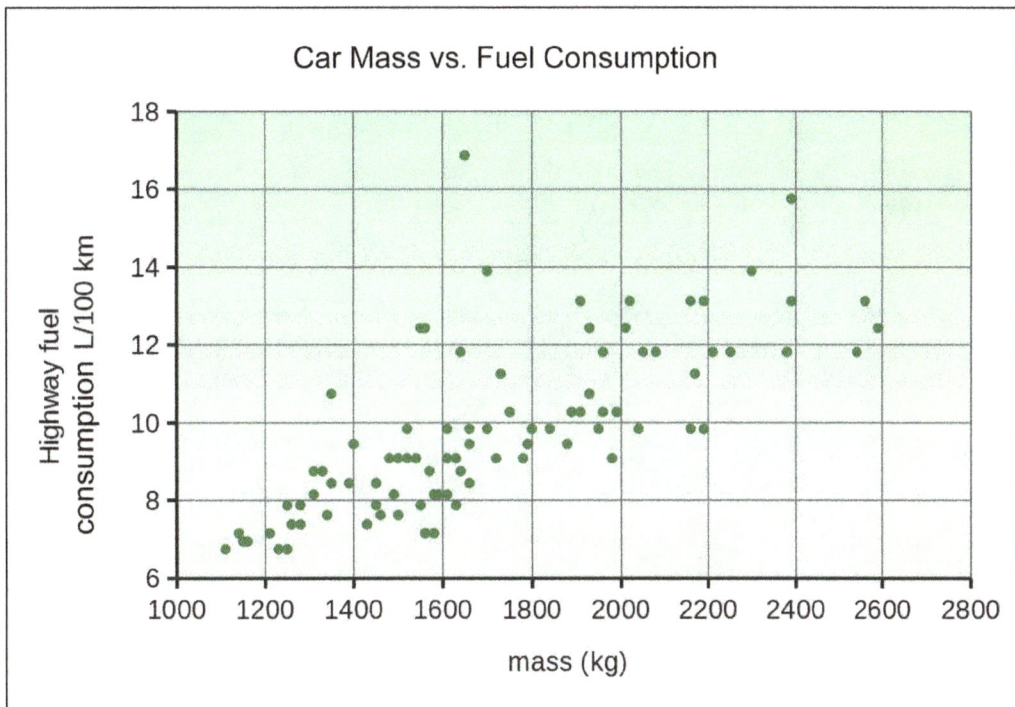

Car Mass vs. Fuel Consumption

a. Describe the general pattern and any special features of the plot.
   Do heavier cars use more or less fuel, in general, than lighter cars?

b. Find the heaviest car that uses less than 10 L per 100 km. How much does it weigh?

c. List the mass and the fuel consumption of the car that is so different from the others that it may be even an error in the data, such as a typo.

d. Among cars weighing from 1800 to 2200 kg, how much fuel does the car with the least fuel consumption use?

3. The data shows the height and the shoe size of a group of individuals. The shoe size was measured in centimetres, to the nearest half centimetre, instead of using shoe size, because shoe sizes vary quite a bit between different brands and countries.

Make a scatter plot. Choose the scaling for both axes wisely. You don't want the data points scrunched up in one area of the graph and most of it being empty. The data points should "fill" the graph area, the best possible, so that the scatter plot features will be well visible.

In this case, you don't want to start the scaling from 0, for either variable. Check the minimum and the maximum values (for both variables), and use those to guide you in regards to scaling.

| Height (cm) | Shoe size (cm) |
|---|---|
| 155 | 23 |
| 158 | 23.5 |
| 160 | 22.5 |
| 160 | 24.5 |
| 162 | 23.5 |
| 162 | 24 |
| 164 | 24 |

| Height (cm) | Shoe size (cm) |
|---|---|
| 165 | 23 |
| 165 | 24.5 |
| 167 | 24 |
| 167 | 25 |
| 167 | 25.5 |
| 168 | 26 |
| 168 | 26 |

| Height (cm) | Shoe size (cm) |
|---|---|
| 170 | 27 |
| 171 | 26 |
| 173 | 26.5 |
| 175 | 26 |
| 177 | 25.5 |
| 178 | 28 |
| 179 | 27.5 |

| Height (cm) | Shoe size (cm) |
|---|---|
| 180 | 28 |
| 182 | 27 |
| 185 | 28 |
| 186 | 29 |
| 186 | 29 |
| 188 | 29.5 |

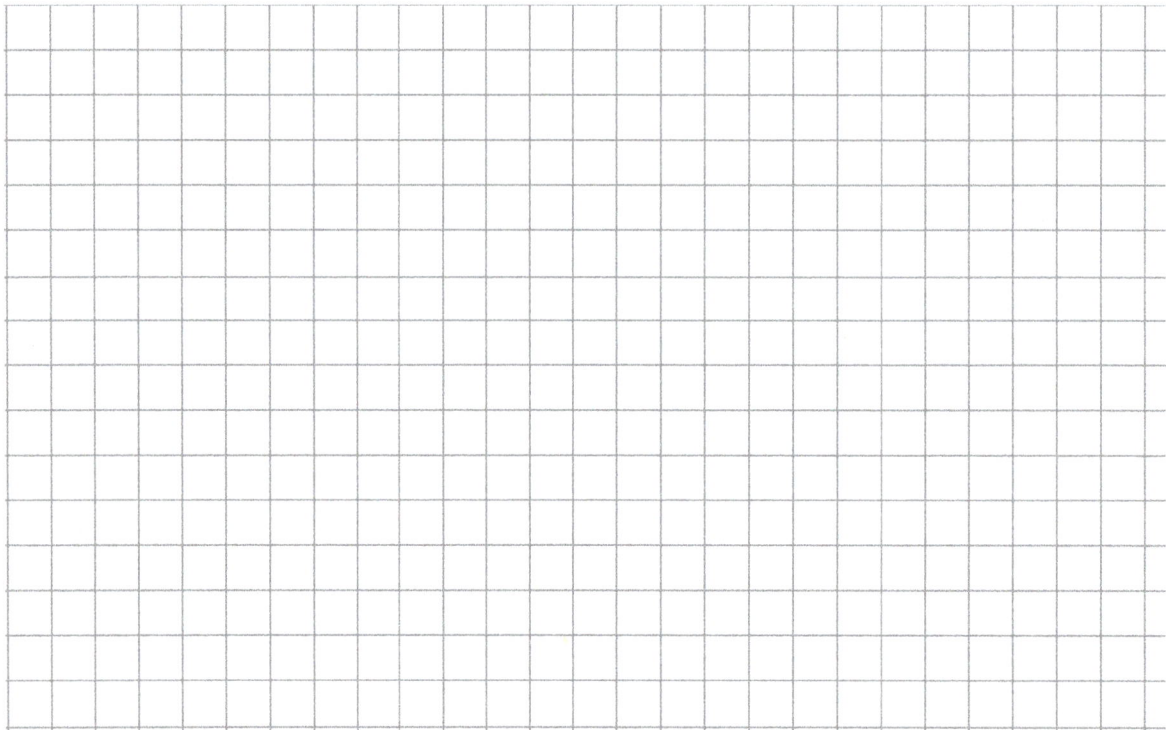

Lastly, describe the basic pattern and possible special features of the scatter plot.

4. The data below gives the weekly work hours and the hourly pay of a company's employees.

Make a scatter plot. Choose the scaling for the axes in such a manner that the data points will "fill" the graph area, the best possible. This means that the scaling may not start from 0. Check what the minimum and the maximum values are, and use those to guide you with the scaling.

| Weekly work hours | 25 | 27 | 28 | 30 | 34 | 34 | 35 | 34 | 35 | 36 | 36 | 36 | 37 | 38 | 38 | 38 | 45 | 45 | 47 | 52 | 52 | 54 | 55 | 59 | 64 |
|---|---|---|---|---|---|---|---|---|---|---|---|---|---|---|---|---|---|---|---|---|---|---|---|---|---|
| Hourly pay ($) | 22 | 15 | 41 | 45 | 36 | 48 | 18 | 20 | 20 | 20 | 21 | 22 | 21 | 20 | 20 | 24 | 32 | 26 | 45 | 35 | 47 | 60 | 22 | 28 | 45 |

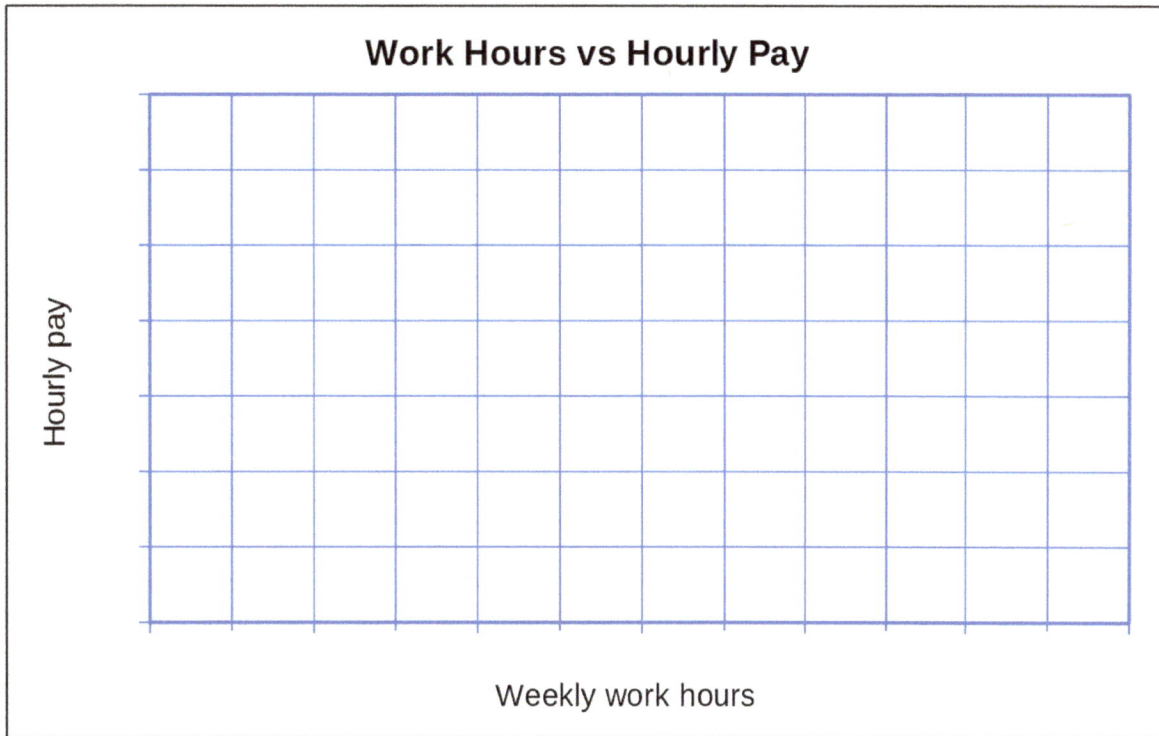

**Work Hours vs Hourly Pay**

Hourly pay

Weekly work hours

5. Refer to the scatter plot you made in question 4 above.

   **a.** Is there a relationship or an association between the two variables? If so, what kind?

   **b.** Describe any special features you see in the scatter plot.

   **c.** Find the person with the highest hourly pay. How many people worked more hours than them?

   How many hours did those people work in a week?

   **d.** How much more did the one who worked the most hours earn than the one who worked the least hours?

   **e.** Did the person with the lowest hourly rate work the least hours? How many hours did they work?

# Fitting a Line

Sofia recorded the distance she walked and the time she took on her daily walks using a pedometer. Then she made a scatter plot from the data:

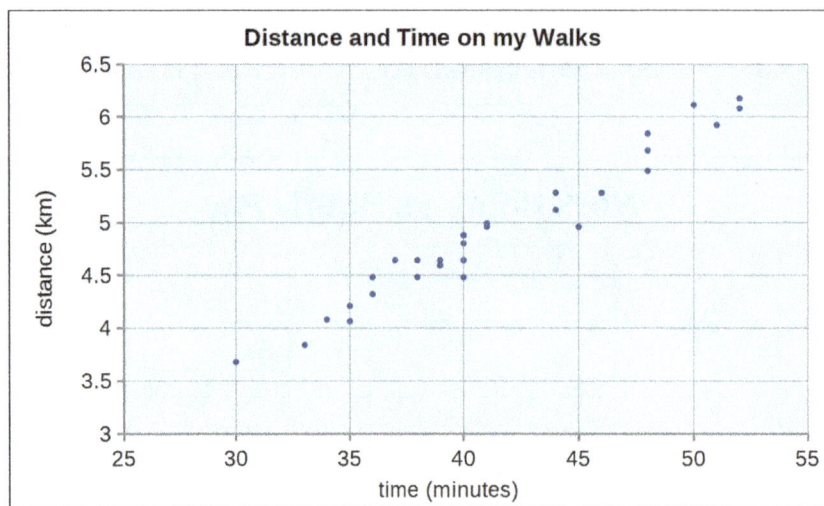

We can see a positive linear association between the variables. It is a common thing to **draw a line to fit the trend in the data** when there is a linear association, so let's look at that.

There is no single right way to do this, not even mathematically. In this lesson, we will not use mathematical algorithms but simply sketch a line in an approximate manner, visually.

Don't attempt to maximize the number of points that the line goes through. Instead, draw a line that will fit the general trend of the data, leaving <u>approximately</u> the same amount of points above the line and below it.

One way is to first sketch an oval that encompasses most of the data points, and then draw a line that is the axis ("diagonal") of the oval.

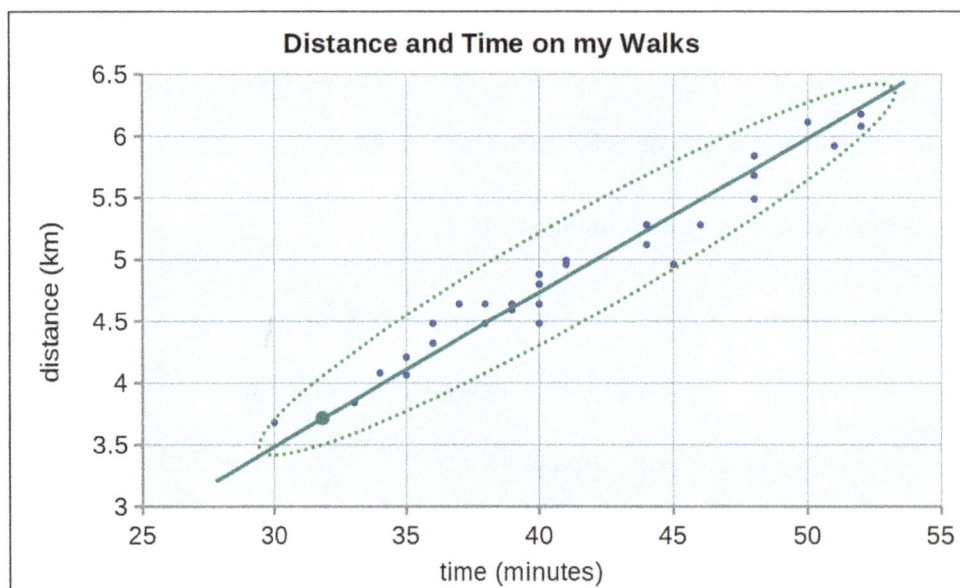

We can then use the line for predicting data values. For example, if Sofia went on a 32-minute walk, it is likely she would walk about 3.7 km (see the turquoise point on the line).

1. **a.** Fit a line to our graph showing the ages of wives and husbands.

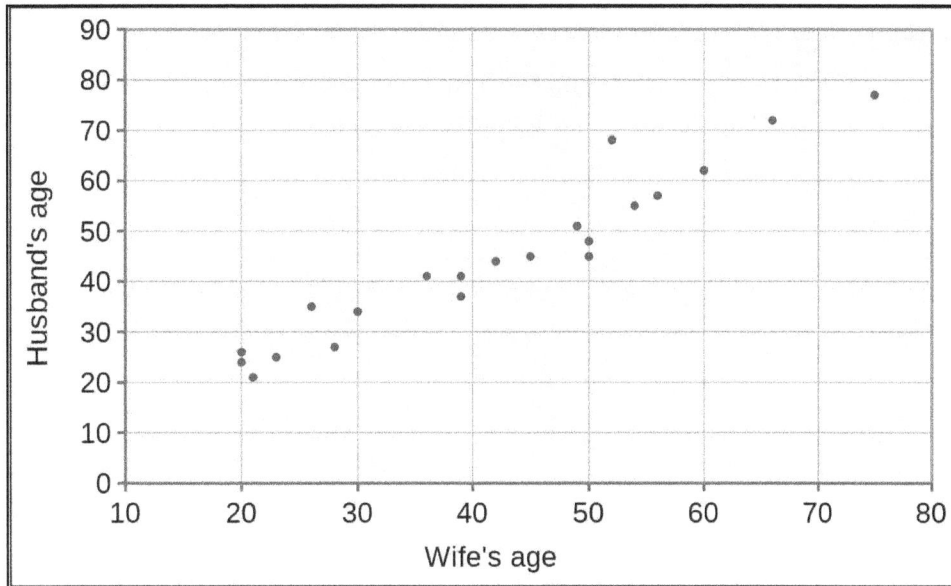

Use your line to answer:

**b.** If the wife is 64 years old, how old would we expect the husband to be?

**c.** For an 80-year-old husband, what age would we expect his wife to be?

2. Andrew drew a line to fit the data in this scatter plot. Notice: while there are five points above the line and seven below it (close to the same number), the points above the line are much closer to the line than the points below it. This is therefore not an accurate fit.

Draw a line that fits the trend in the data better. Think, not only about the number of points, but also about the distances of those points to the line, trying to "equalize" the total of those distances on each side of the line.

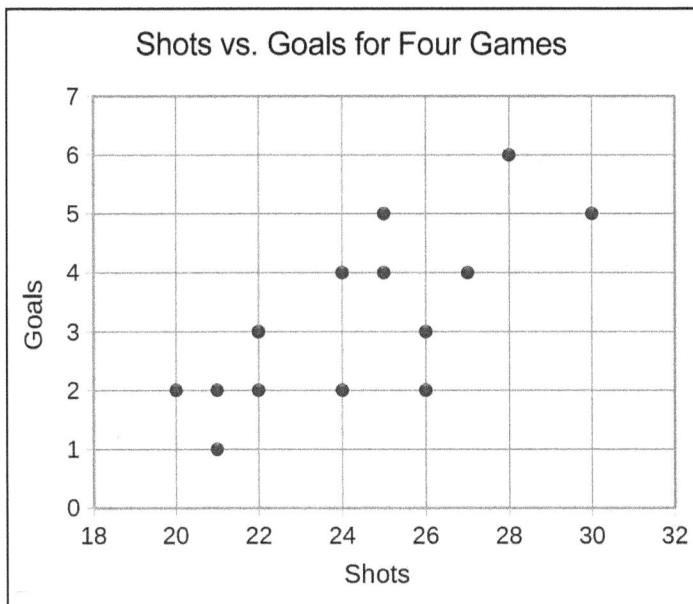

3. The scatter plot below shows the life expectancy at birth versus the birth rate per 1000 people of various Asian countries.

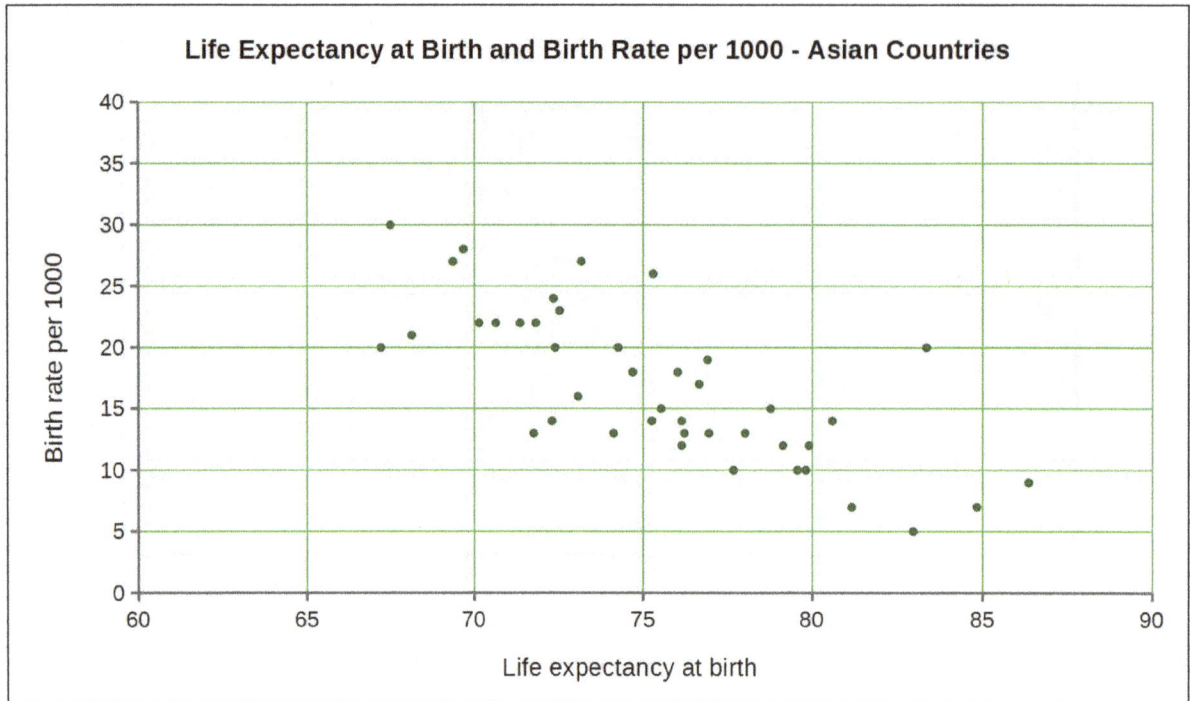

**a.** Draw a line to represent the trend in the data. Assess the fit of your line by the closeness of the data points to the line.

**b.** Using your line, what would you predict the life expectancy to be for a country with a birth rate of 20 per 1000?

**c.** Find the country (find its point) with a birth rate of 30 per 1000. What is its life expectancy at birth?

What is the predicted life expectancy for such a country based on the line you drew?

**d.** What is the difference between the predicted and the actual life expectancy for the country with the lowest birth rate in this sample?

**e.** What is the difference between the predicted and the actual birth rate per 1000 for the country with the highest life expectancy in this sample?

4. **a.** Draw a line that best represents the trend in the data in the graph below. Judge the fit of your line by checking how close the data points (on both sides) are to the line.

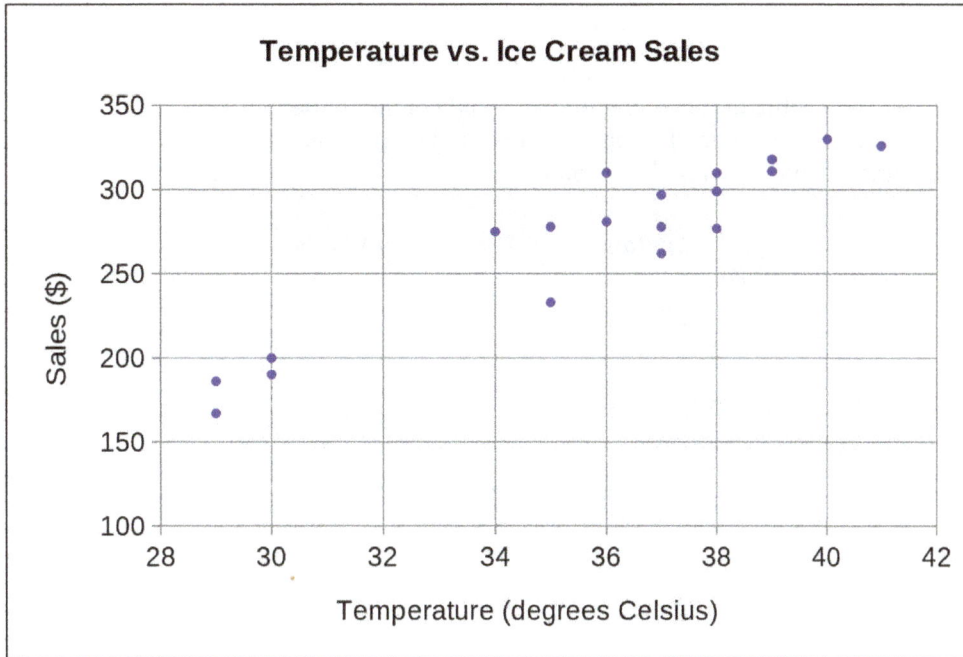

**b.** Predict what the ice cream sales would be if the temperature was 32°.

5. **a.** Draw a line that you feel best fits the data in the scatter plot below.

**b.** Why is it not a good idea to simply draw a line from point A to point B?

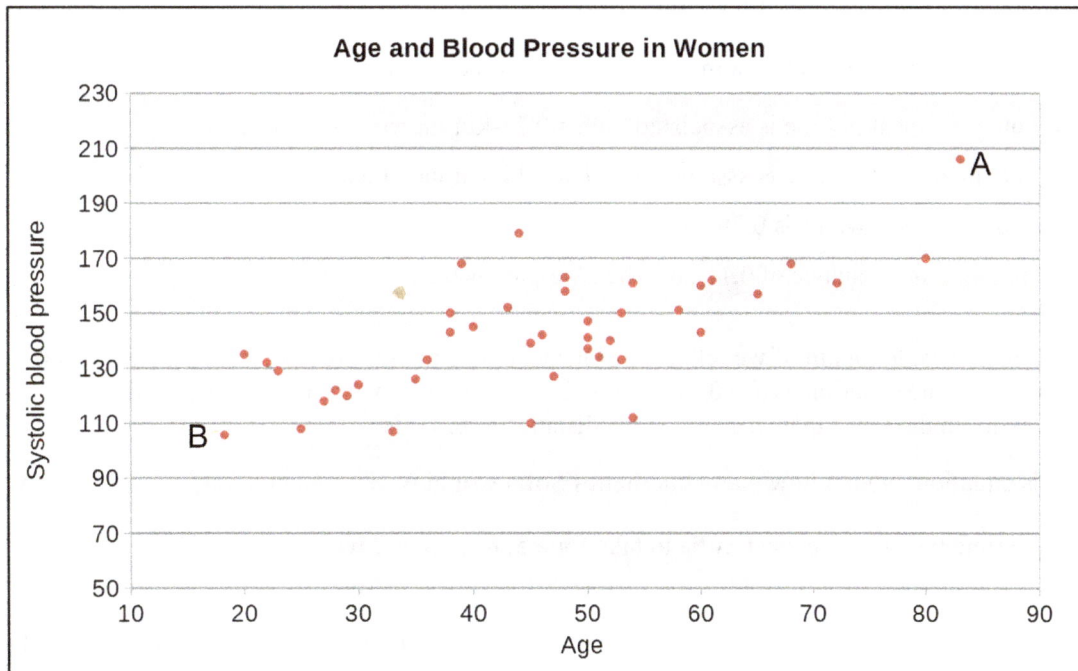

# Equation of the Trend Line

Going back to the Sofia's walks, the graph shows a line that is fitted to the data points. Its equation is $d = 0.11t + 0.28$, where $d$ is the distance in kilometres and $t$ is the time in minutes. The equation was calculated by a spreadsheet program. The line we drew in the previous lesson by using the ellipse method is quite close to this line calculated with an algorithm.

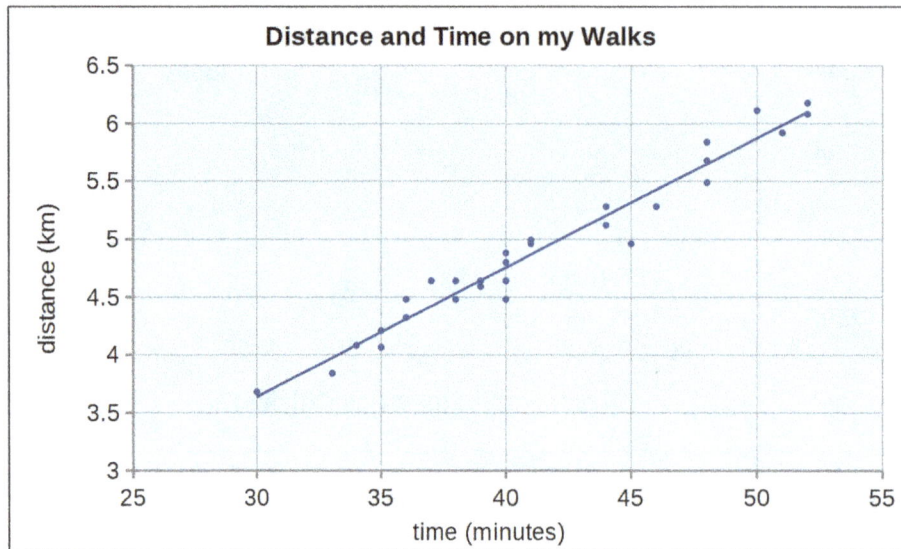

**Distance and Time on my Walks**

*distance (km)* — *time (minutes)*

The slope of the line, 0.11, tells us that for each one-minute increment, the distance increases by 0.11 km. In other words, Sofia tends to walk at a speed of about 0.11 km per minute.

The $d$-intercept, 0.28 km, means that the equation predicts that at zero minutes, Sofia would have walked 0.28 km. This doesn't make sense. Keep in mind that the equation is calculated using data that was limited to between 30 and 52 minutes. Estimating values outside of that range is called extrapolating, and we need to be careful with that. The equation is a model, and it may not be valid outside the original range of values.

1. Use the equation, $d = 0.11t + 0.28$ as a model for Sofia's walks. Choose all the correct statements.

   - Each 1-minute increment in time is associated with a 0.28-km increment in distance.

   - Each 1-minute increment in time is associated with a 0.11-km increment in distance.

   - The average distance she walks is 0.28 km.

   - The equation predicts a distance of 0.11 km when the time is 0.28 minutes.

2. Concerning Sofia's walks again, if we tell the spreadsheet program to force the line to go through (0, 0), the program calculates the equation as $d = 0.119t$. This line is a slightly worse fit considering the data points, but it matches reality in the sense that when $t = 0$, the distance is zero also.

   **a.** Using this equation, predict how many kilometres Sofia would walk in 32 minutes.

   **b.** How much time would you expect Sofia to take for a 5.3-kilometre walk?

   **c.** Use the equation $d = 0.119t$ and fill in. Each five minutes added to the walk time is associated with

   a _____-km increase in her walking distance.

3. The scatter plot below shows the weight and height of various adult female German shepherds. (It does not have to do with weight gain/loss of an individual dog — each dot signifies a different dog.) The equation for the trend line is $h = 0.91w + 29.7$, where $h$ is the height in centimetres and $w$ is the weight in kilograms.

**Female German Shepherd Weight and Height**

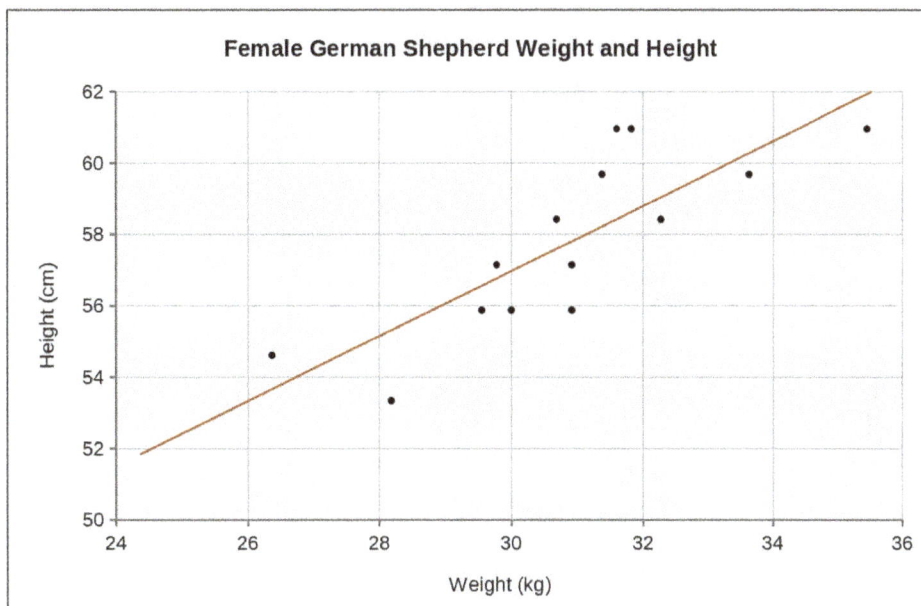

**a.** Which statements are correct?

- Each 0.91 kg increase in weight is associated with a 1 cm increase in height.

- Each 1 kg increase in weight is associated with a 0.91 cm increase in height.

- Heavier dogs tend to be taller; and for each 2 kg increase in the weight, the dogs tend to be 1.8 cm taller.

- The model predicts a height of 29.7 cm for a dog weighing 0 kg.

- The model predicts a weight of 29.7 kg for a dog that is zero cm tall.

- We should be careful in using this model to extrapolate the heights of dogs less than 26 kg.

**b.** Use the equation to predict the weight of a dog that is 57 cm tall, to the nearest kilogram.

**c.** Use the equation to predict how tall a dog weighing 29 kg would be.

**d.** Would a dog that weighs 27 kg and is 53 cm tall be considered an outlier?

**e.** What is the difference between the predicted height of a 34-kg dog and its real height, if in reality it is 62 cm tall?

4. The line fitted to the data about the mass ($m$) and fuel consumption (FC, in litres per 100 km) of cars has the equation

$$FC = 0.004411m + 2.28257$$

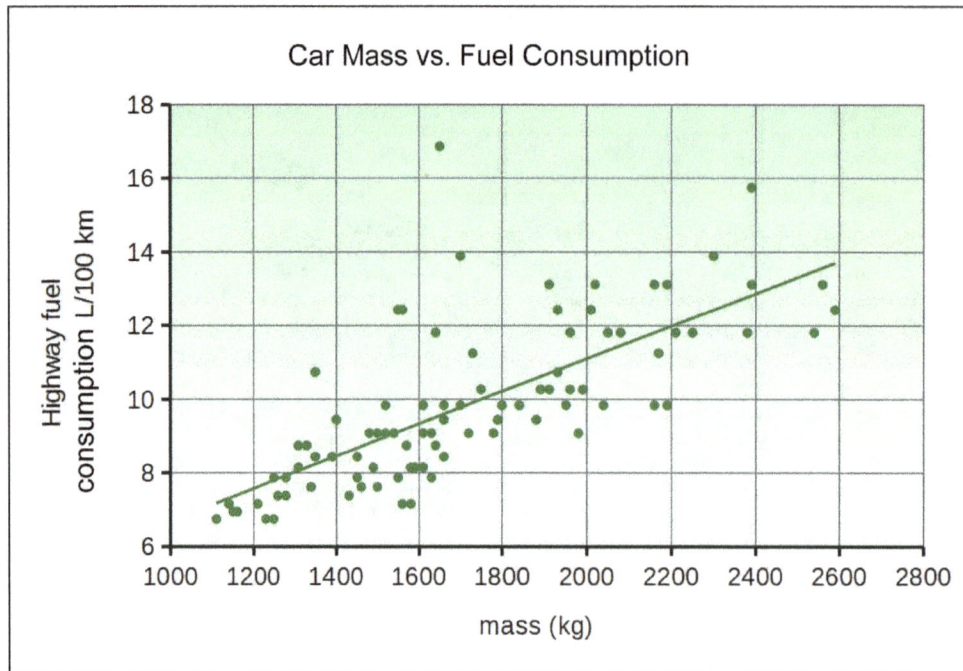

Car Mass vs. Fuel Consumption

**a.** What does the slope of 0.004411 signify in this context?

**b.** For each 100-kg increase in a car's mass, how much would you expect the fuel consumption to change?

**c.** Predict the fuel consumption of a 2500-kg car.

**d.** If a car uses 8.6 litres/100 km in highway driving, what would you expect its mass to be?

**e.** Find the dot in the graph for Lamborghini Murcielago A-S6, which weighs 1650 kg and uses 16.875 litres (a lot!) per 100 km in highway driving. Based on the equation of trend line, what would you expect this car's fuel consumption to be?

**f.** Let's say you want to buy a car that uses at most 9 litres per 100 km in highway driving. Based on the scatter plot, how much would you expect your car to weigh, at a maximum?

5. **a.** Draw a line to fit the trend in the scatter plot below.

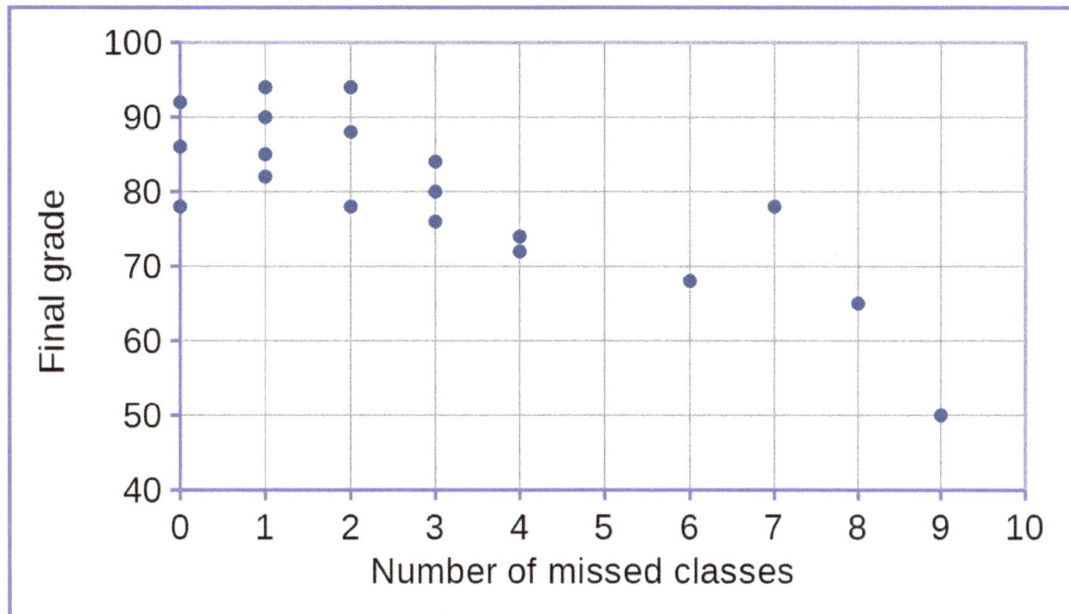

**b.** Find the (approximate) equation of your line.
   *Hint: find the y-intercept and the slope.*

**c.** What does the slope of your equation signify in this context?

**d.** What does the *y*-intercept of your equation signify in this context?

**e.** Using your equation, predict what the final grade would be for a student that missed five classes.

**f.** Tanya calculated the equation of the trend line as $G = 3m + 89$, where G is the grade and *m* is the number of the missed classes. Explain why this cannot possibly fit the data.

6. The data shows shots the attempted shots and the shots made of 18 basketball players.

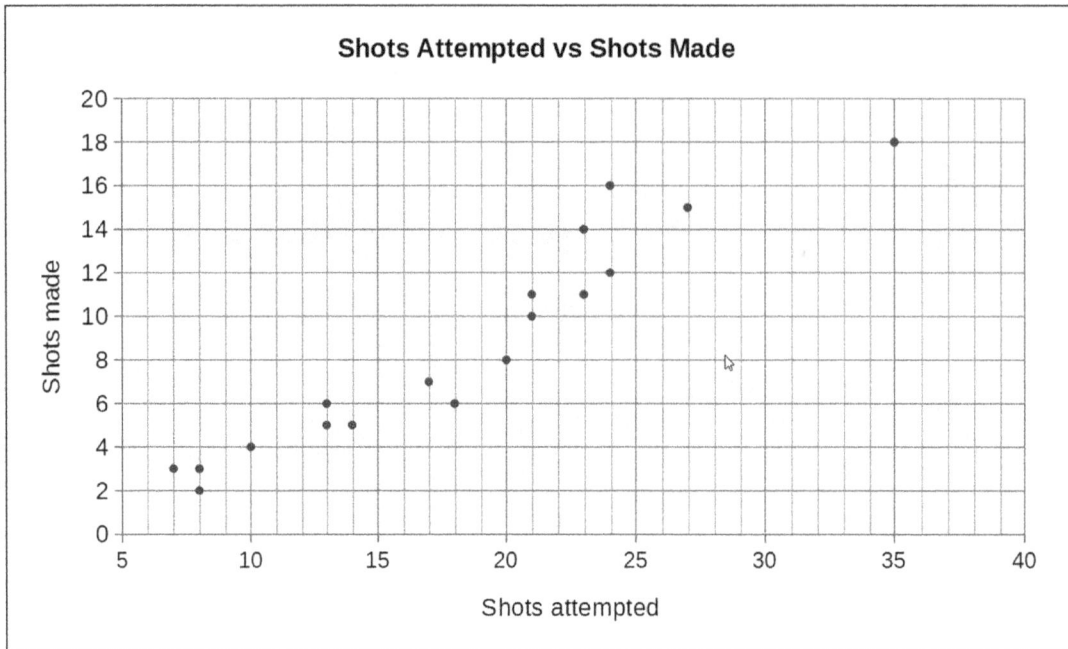

**Shots Attempted vs Shots Made**

**a.** Draw a line that fits the trend in the data.

**b.** Find the equation of your line.

**c.** What does the slope in the equation signify in this context?

**d.** And the *y*-intercept?

**e.** If a player attempts 30 shots, how many shots does your model predict they would make?

**f.** What prediction does your model make about the number of shots attempted by a player if they made 8 shots?

# Two-Way Tables

We have been studying bivariate data, in other words, data that involves two variables. So far, that data has all been numerical: we have had two numerical values for each data item, and therefore, we have been able to plot the data as points in the coordinate plane, the two coordinates being the values of the two variables.

Now we will look at bivariate data that is organized into categories, and the values may or may not be numerical.

The **two-way table** on the right shows how many students in each grade of an elementary school can swim, and how many cannot. It also shows the row and column totals.

It is called a two-way table because it records the information from *two* variables. In this case, the two variables are the student's grade level and whether the student can swim or not.

The first variable takes numerical values from 1 to 5. The table has a row corresponding for each of those values. The second takes the values "Yes" and "No", and those correspond to the two columns labelled "can swim" and "cannot swim".

| Grade | can swim | cannot swim | Total |
|---|---|---|---|
| 1 | 18 | 30 | 48 |
| 2 | 26 | 25 | 51 |
| 3 | 38 | 12 | 50 |
| 4 | 44 | 5 | 49 |
| 5 | 48 | 4 | 52 |
| **TOTAL** | **174** | **76** | **250** |

The original data may look like this: (1, no), (1, yes), (2, no), (2, no), (3, no), (5, yes), etcetera, each pair of data depicting one student. But we cannot analyze or summarize the data when it is in that format. A two-way table allows us to **categorize and tally up the data**, and then also to analyze it to see if there is any **association between the two variables**.

In this case, we can see that with advancing grade level, there are many more students who can swim than who cannot. In 5th grade, nearly all students can swim. So, there *is* an association between the variables.

The data from a two-way table can be presented as a double-bar graph (left), or a stacked bar graph (right):

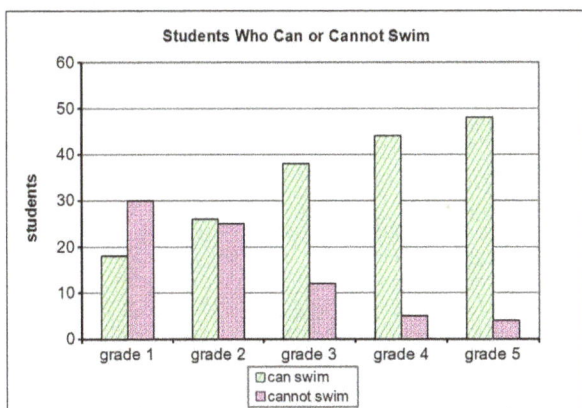

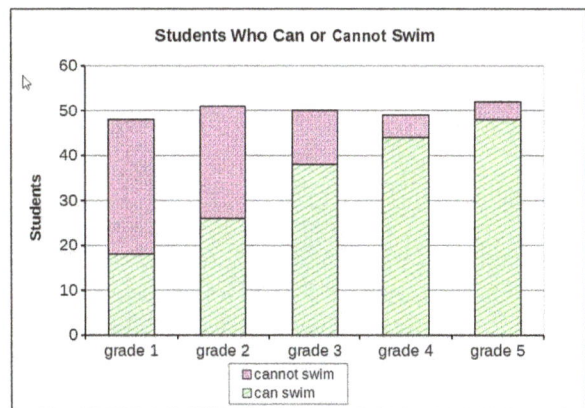

A stacked bar graph is more common. From the graphs, it is easy to see that as the grade level advances, more and more students are able to swim.

1. A community college tracked how many of their students in a given year completed a certain course that was offered both as an in-person course and as an online version.

| | completed the course | did not complete the course | Total |
|---|---|---|---|
| In person | 57 | 8 | 65 |
| Online | 23 | 25 | 48 |
| **TOTAL** | 80 | 33 | 113 |

**a.** Looking only at those who took the online course, what percentage of them completed the course? What percentage did not?

**b.** Looking only at those who took the in-person course, what percentage of them completed the course? What percentage did not?

**c.** Is there a relationship or association between the variables? Explain.

2. Jordan asked a bunch of people at a local gym as to their opinion on increasing the membership price in return for some improved facilities. He categorized the people as 20- to 30-year-olds and as 31+ year-olds. Here are some highlights of Jordan's research:

- Of the seventy-seven 20- to 30-year-olds, 21 were in favor of the plan.

- In total, there were 79 people for and 70 people against this plan.

**a.** Fill in the two-way table below from this data. Use the two age groups as rows. Fill in all the numbers.

| | for | against | Total |
|---|---|---|---|
| 20-30 years old | | | |
| 31+ years old | | | |
| **TOTAL** | | | |

**b.** Overall, are most of the 20- to 30-year-olds for or against this plan?

What about of the 31+ year-olds?

**c.** Now look at the column totals. Are there more people in general that are for the price increase or that are against it?

**d.** Do you feel there is an association between the variables? Explain.

**e.** Why do you think so many of the younger people are against this plan?

3. At a family reunion, Ashley asked all her relatives whether at the next year's reunion they should pay a fee and gather at the local amusement park grounds. Here are the answers she got.

- Adults: (18) no, no, no, yes, no, no, yes, no, no, yes, yes, no, no, no, no, no, yes, yes

- Teens: (8) yes, yes, yes, yes, yes, yes, no, no

- Children (16): yes, yes, yes, yes, yes, yes, yes, yes, yes, yes, yes, yes, yes, yes, yes, yes

**a.** Create a two-way table from the results.

**b.** Based on the results, are there more people in favor or against gathering at the amusement park next year?

**c.** Does the answer change if children are ignored?

**d.** What percentage of the adults are in favor of this?

What percentage of the teens?     Of the children?     Of everyone?

**e.** The two variables are: age group (adult/teen/child) and opinion (yes/no). Is there a relationship or an association between the variables? What kind?

---

## Puzzle Corner

Is there an association between the variables? Explain.

**Type of schooling and number of pets**

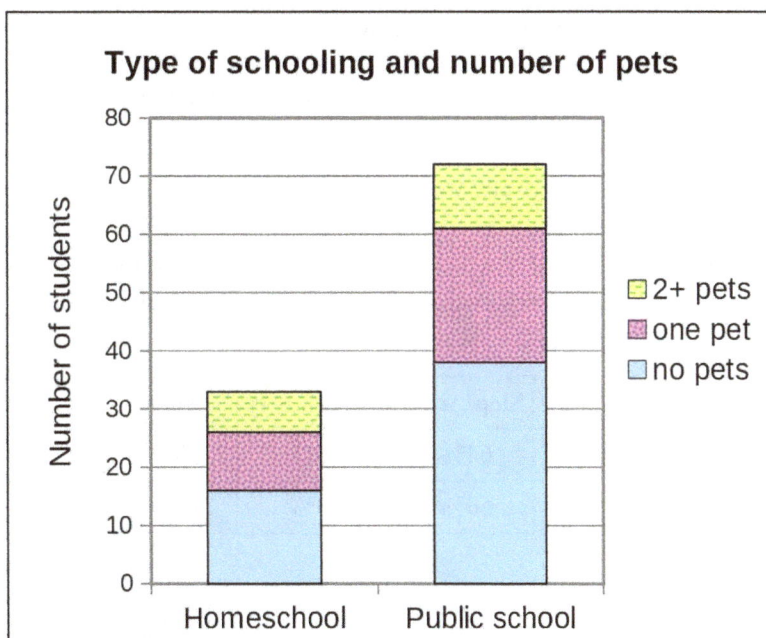

# Relative Frequencies

**Example 1.** Jeanine wanted to find out if walking affected her sleep quality, so, for 60 days, she recorded whether she took a walk or not, and how she slept the following night. The results are below:

|  | Slept well | Slept poorly | Total |
|---|---|---|---|
| Walked | 29 | 14 | 43 |
| Didn't walk | 4 | 13 | 17 |
| **TOTAL** | 33 | 27 | 60 |

If you look at the "Slept poorly" column only (14 vs. 13), it would appear that walking has no effect. However, we need to notice the fact that there were far more days she took a walk than days that she didn't.

Enter percentages! Let's calculate the **relative frequencies** (as fractions and percentages) for each row, using the <u>row</u> totals. To do that, we simply write the fraction, and then use a calculator to get a percentage. (Here, it does not make sense to include a row listing the column totals, because we're using the *row* totals to calculate the relative frequencies, so that row is omitted).

|  | Slept well | Slept poorly | Total |
|---|---|---|---|
| Walked | 29/43 = 67.4% | 14/43 = 32.6% | 100% |
| Didn't walk | 4/17 = 23.5% | 13/17 = 76.5% | 100% |

Let's now look at the table that shows *only* the percentages, so we can more clearly see whether there is an association:

|  | Slept well | Slept poorly | Total |
|---|---|---|---|
| Walked | 67.4% | 32.6% | 100% |
| Didn't walk | 23.5% | 76.5% | 100% |

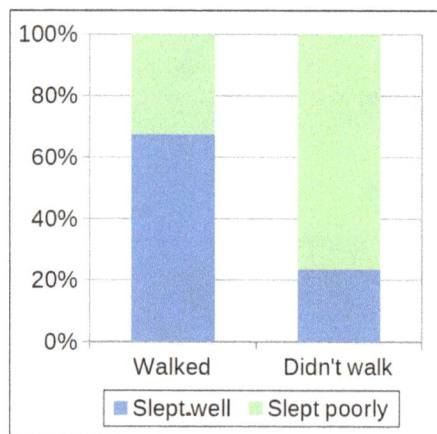

Now it is easy to see that when she took a walk, most of the time she did sleep well, and when she didn't take a walk, most of the time she did not sleep well. So, there is an association!

The stacked bar graph also shows the same: she sleeps better much more often when she walks.

Here is an example of how the relative frequencies could look like **IF** there was no association. The percentages in each row would be similar:

|  | Slept well | Slept poorly | Total |
|---|---|---|---|
| Walked | 64% | 36% | 100% |
| Didn't walk | 60% | 40% | 100% |

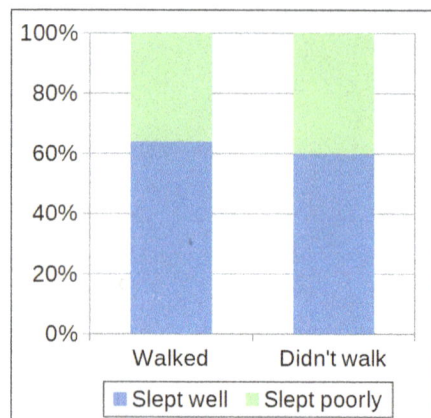

1. **a.** Calculate the relative frequencies to the nearest tenth of a percent. Write them in the empty table.

| | Does crossword puzzles frequently | | |
|---|---|---|---|
| | Yes | No | Total |
| 20-39 years | 17 | 55 | 72 |
| 40-59 years | 68 | 90 | 158 |
| 60-79 years | 64 | 56 | 120 |
| TOTAL | 149 | 201 | 350 |

| | Does crossword puzzles frequently | | |
|---|---|---|---|
| | Yes | No | Total |
| 20-39 years | | | |
| 40-59 years | | | |
| 60-79 years | | | |

**b.** Which age group has the largest proportion of crossword puzzle fans?

**c.** Does there appear to be an association between age and being a crossword puzzle fan? Explain and support your answer.

2. **a.** Calculate the relative frequencies to the nearest tenth of a percent using the row totals. This time, include them in the same table as the numbers.

| | Owns a house or an apartment | | | | | |
|---|---|---|---|---|---|---|
| | Yes | | No | | Total | |
| | # | % | # | % | # | % |
| Has debt | 274 | 42.5% | 371 | | 645 | |
| Has no debt | 129 | | 160 | | 289 | |
| TOTAL | 403 | | 531 | | 934 | |

**b.** Would you say there is an association between having debt, and owning a house/apartment? Explain.

**c.** Look at the *column* for those who own a house/apartment. If you randomly pick a person who owns a house/apartment, is it more likely that they do have debt or that they don't?

**d.** Look at the column for those who don't own a house or an apartment. If you randomly pick one such person, is it more likely that they do have debt or that they don't?

**e.** Now look at the row of those who have debt. If you randomly choose a person who has debt, is it more likely that they own a house/apartment or that they don't?

3. Mike owns a restaurant. He was interested in knowing whether customers who buy the expensive dishes also buy dessert more often than the customers who buy cheaper dishes. He recorded his customers' purchases over one day; that data is listed below. Your task is to make a two-way table from the data, calculate the relative frequencies, and check if there is an association or not.

Each entry below pertains to one customer. Dishes 1 and 2 are expensive, and the rest are not.

| | | |
|---|---|---|
| Dish 1/no dessert | Dish 3/no dessert | Dish 4/dessert |
| Dish 1/no dessert | Dish 3/no dessert | Dish 4/dessert |
| Dish 2/no dessert | Dish 3/no dessert | Dish 5/no dessert |
| Dish 2/no dessert | Dish 3/dessert | Dish 5/no dessert |
| Dish 2/no dessert | Dish 3/dessert | Dish 5/no dessert |
| Dish 2/no dessert | Dish 4/no dessert | Dish 5/dessert |
| Dish 1/dessert | Dish 4/no dessert | Dish 5/dessert |
| Dish 1/dessert | Dish 4/no dessert | Dish 5/dessert |
| Dish 1/dessert | Dish 4/no dessert | Dish 6/no dessert |
| Dish 2/dessert | Dish 4/no dessert | Dish 6/no dessert |
| Dish 2/dessert | Dish 4/dessert | Dish 6/dessert |
| Dish 2/dessert | Dish 4/dessert | Dish 6/dessert |
| Dish 2/dessert | Dish 4/dessert | Dish 6/dessert |
| Dish 2/dessert | Dish 4/dessert | Dish 6/dessert |

a. Make a two-way table, using two categories for the dish: either expensive or cheap, and using, again two categories for the dessert: yes or no. Then calculate the relative frequencies. You can make a separate table for those like was done in example 1, or include them in the same table, like in exercise 2.

b. Is there an association between the variables (price of dish and getting a dessert or not)? Justify your answer.

c. Based on this data, should Mike change the layout of the menu, and move the dessert section fairly close to the section of the expensive dishes?

4. The table below shows the number of teenagers who had had a summer job the previous summer.

**a.** Ryan says there is no association because the numbers 33 and 35 are very close, and also the numbers 55 and 53. To answer him, first calculate the relative frequencies based on the row totals.

|  | Had a summer job | | |
|---|---|---|---|
|  | Yes | No | Total |
| 14-15 years | 33 | 109 |  |
| 16-17 years | 35 | 55 |  |
| 18-19 years | 68 | 53 |  |
| TOTALS |  |  |  |

Relative frequencies:

|  | Had a summer job | | |
|---|---|---|---|
|  | Yes | No | Total |
| 14-15 years |  |  |  |
| 16-17 years |  |  |  |
| 18-19 years |  |  |  |

**b.** Then make a stacked bar graph of the data. Use two different colours or two different patterns to colour the bars, and fill in the legend.

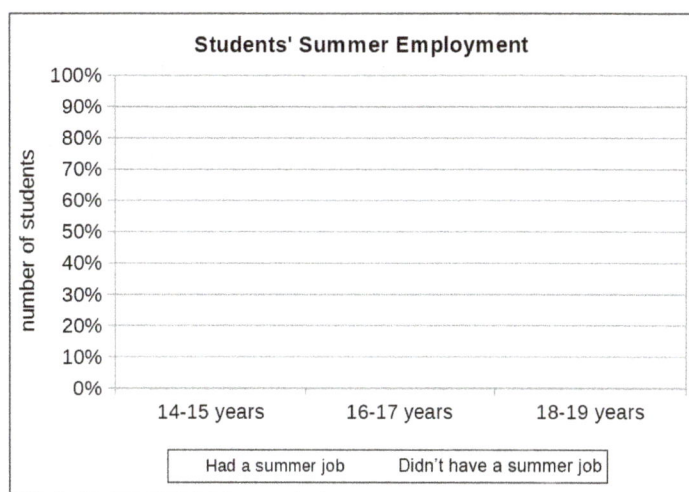

Students' Summer Employment

number of students — 100%, 90%, 80%, 70%, 60%, 50%, 40%, 30%, 20%, 10%, 0%

14-15 years    16-17 years    18-19 years

Had a summer job    Didn't have a summer job

**c.** Based on the data and the graph, what would you say to Ryan? Be convincing — support your answer!

**d.** Which statements below are true? Choose all that are true.

(i) Approximately the same number of people were surveyed from each of the three age groups.

(ii) About 2/3 of all the surveyed teenagers had had a summer job.

(iii) About 3/4 of the youngest age group had not had a summer job.

(iv) About 2/5 of all the surveyed teens had had a summer job.

5. A researcher asked some people over 50 years of age and some people between 30 and 50 years of age about their favorite type of books. The results are in the table below, on the left.

**Favourite book genres by age**

|  | 30-50 | over 50 | Total |
|---|---|---|---|
| Biographies & Memoirs | 36 | 245 | 281 |
| Comics | 126 | 45 | 172 |
| Mysteries | 186 | 357 | 543 |
| Poetry | 22 | 56 | 78 |
| Romance | 215 | 128 | 343 |
| Science Fiction & Fantasy | 267 | 37 | 304 |
| **Total** | **852** | **868** | **1720** |

**Favourite book genres by age**

|  | 30-50 | over 50 |
|---|---|---|
| Biographies & Memoirs | 4.2% | |
| Comics | | |
| Mysteries | | |
| Poetry | | |
| Romance | | |
| Science Fiction & Fantasy | | |
| **Total** | | |

**a.** Calculate the relative frequencies <u>for each column</u> to the nearest tenth of a percent, and write them in the table above, on the right. This means you will use **the column totals** to calculate them. For example, for 30-50 year olds, the percentage for biographies & memoirs is 36/852 ≈ 4.2%.

**b.** List the three most popular genres for each age group.

- 30-50 year olds:

- over 50 year olds:

**c.** What else can you notice? Make at least three other observations about the data.

**d.** Is there an association between the variables? Explain why you feel that way.

6. (optional) Collect data from some population, observing two variables. Present your results in a two-way table, and use relative frequencies to determine whether there is an association between the variables.

For example, you could ask a group of people something about their likes, and categorize the people by age, gender, or something else. As one idea, one variable could be gender and the other could be the person's favourite type of pastime, snack, dessert, colour, etcetera.

Or, the population you're observing could be days instead of people. On each day, you would observe two variables, such as whether it is cloudy or clear in the morning and what time you or some other person wakes up. There are lots of possibilities! The main thing is that the values of the variables need to be in categories. If the values are numerical (such as age), you need to group or categorize them. If both variables are numerical, making a scatter plot may be a better approach than a two-way table.

# Mixed Review Chapter 8

1. This is a triangular prism (a tent-shaped structure).
   Find the length marked with $x$.

   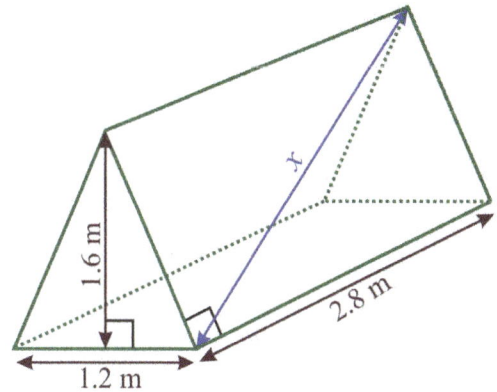

2. Find the statements that are in error, and correct them.

   **a.** $\dfrac{79}{123}$ is rational because it is a fraction (a whole number divided by a whole number).

   **b.** $\sqrt{121}$ is irrational because it has a square root.

   **c.** $3\pi$ is irrational because $\pi$ is irrational, and an irrational number multiplied by a rational number (3) is irrational.

   **d.** 0.19191919... is irrational because it is an unending decimal.

3. Find the (principal) square roots.

   **a.** $\sqrt{10^2 - 6^2}$        **b.** $\sqrt{29 + 7}$        **c.** $\sqrt{\dfrac{81}{16}}$

   **d.** $\sqrt{0.16}$        **e.** $\sqrt{0.01}$        **f.** $\sqrt{1.21}$

4. Solve the equations. Give each solution rounded to three decimals.

| a. $18 + t^3 = 134$ | b. $5x^2 = 220$ | c. $109 - v^3 = 11$ |
|---|---|---|
| | | |

5. Calculate the length of the hypotenuse of a right triangle if its two legs measure 2.31 m and 2.80 m.

6. Lines *l* and *m* intersect at point A.
   Find the measure of angle *x*.

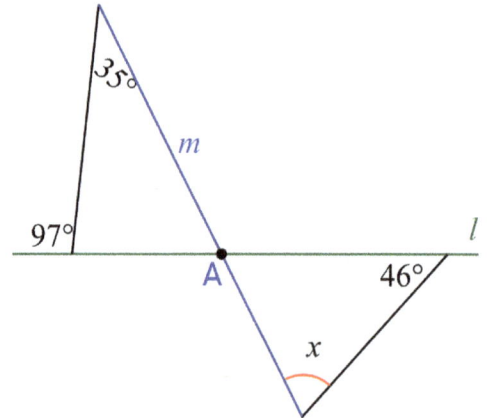

7. Graph the volume of a water reservoir as a function of time to match the description below.

   Your graph will be a rough sketch only, without specific scale on either axis.

   From June onward, the level of water was decreasing in a linear manner, over the summer months. During September, the level of water increased continually, quickly at first, and then at a slower and slower pace, because of rains. For the month of October, there was no change in the water level. The level then increased continually in a linear manner through November and December. In January, there again was no change.

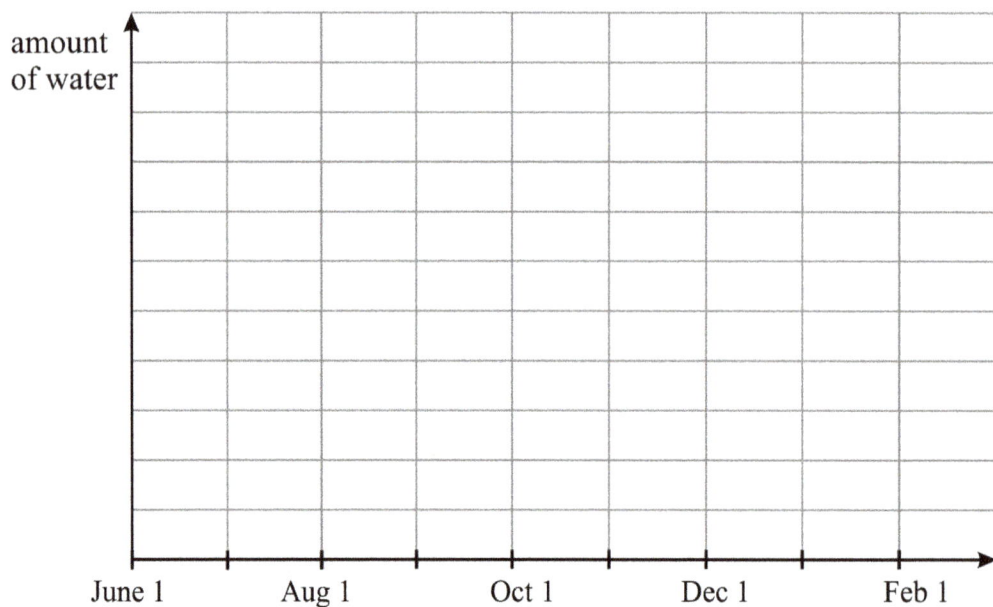

8. **a.** Line L is plotted on the right. Line M is parallel to line L, and passes through the point $(3, 0)$.

Find the equation of Line M, in slope-intercept form, and plot it.

**b.** Line N is perpendicular to line L, and passes through the point $(3, 0)$.

Find the equation of Line N, in slope-intercept form, and plot it.

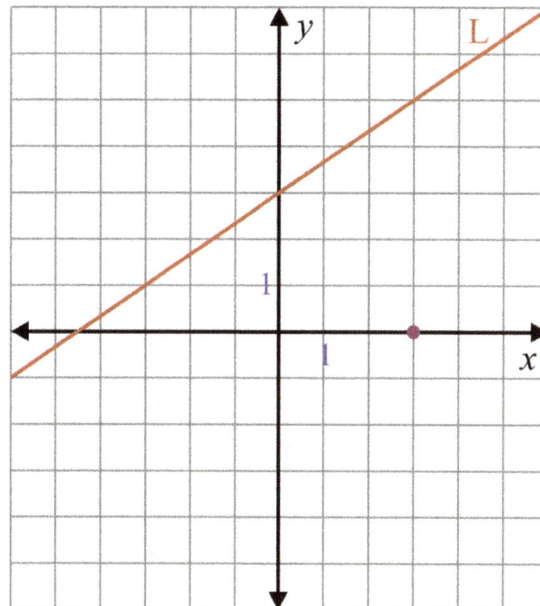

9. Solve each system of equations.

**a.** $\begin{cases} 5x - 7y = 3 \\ -6x + 8y = 1 \end{cases}$

**b.** $\begin{cases} 3x = -2(y + 1) \\ -x - 3y = 0 \end{cases}$

10. A group of chickens and cows has 42 heads and 100 legs. How many chickens and how many cows are there?

11. The scatter plot shows the age and the resting heart rate of a group of males who use a fitness wearable.

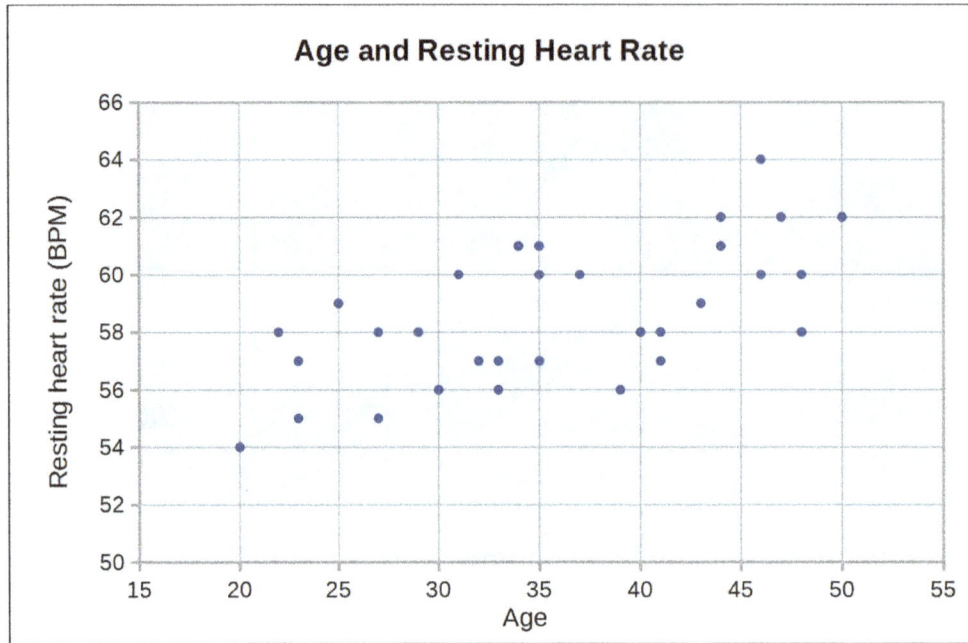

**Age and Resting Heart Rate**

**a.** Describe the general trend and any special features in the scatter plot.

**b.** Draw a trend line on the graph.

**c.** Find the equation of your line.

**d.** According to your equation, when a person's age increases by one year, what change in the resting heart rate is associated with that?

**e.** What heart rate does your equation predict for a 30-year-old male who wears a fitness wearable?

**f.** What heart rate does your equation predict for a newborn?
    (A normal heart rate for newborns is from 120 to 160 beats per minute.)

What does that tell us about extrapolating with this data?

12. Claire owns a salon. She conducted a survey among her clientele concerning how often they had a haircut, on average. The results are shown in the table below.

**Relative frequencies:**

|  | Male | Female | Total |
|---|---|---|---|
| every month | 56 | 15 | 71 |
| every two months | 31 | 37 | 68 |
| every three months | 17 | 46 | 63 |
| every four months | 8 | 25 | 33 |
| TOTALS | 112 | 123 | 235 |

|  | Male | Female |
|---|---|---|
| every month |  |  |
| every two months |  |  |
| every three months |  |  |
| every four months |  |  |
| TOTALS | 100% | 100% |

**a.** Which statement(s) are correct?

   **(i)** Among those who get a haircut every three months, most of them are female.

   **(ii)** If you randomly choose a male client from Claire's clientele, most likely they get a haircut every two months.

   **(iii)** Most of the clients who get a haircut every month are males.

**b.** Calculate the relative frequencies based on the *column* totals, to the nearest percent.

**c.** Is there an association between the variables? Explain.

# Chapter 8 Review

1. The scatter plot below shows the life expectancy at birth and the birth rate per 1000 population for almost 200 countries in the world.

   **a.** Describe the pattern of association you see in the scatter plot.

   **b.** Describe any special features visible in the scatter plot.

**Life Expectancy at Birth and Birth Rate per 1000**

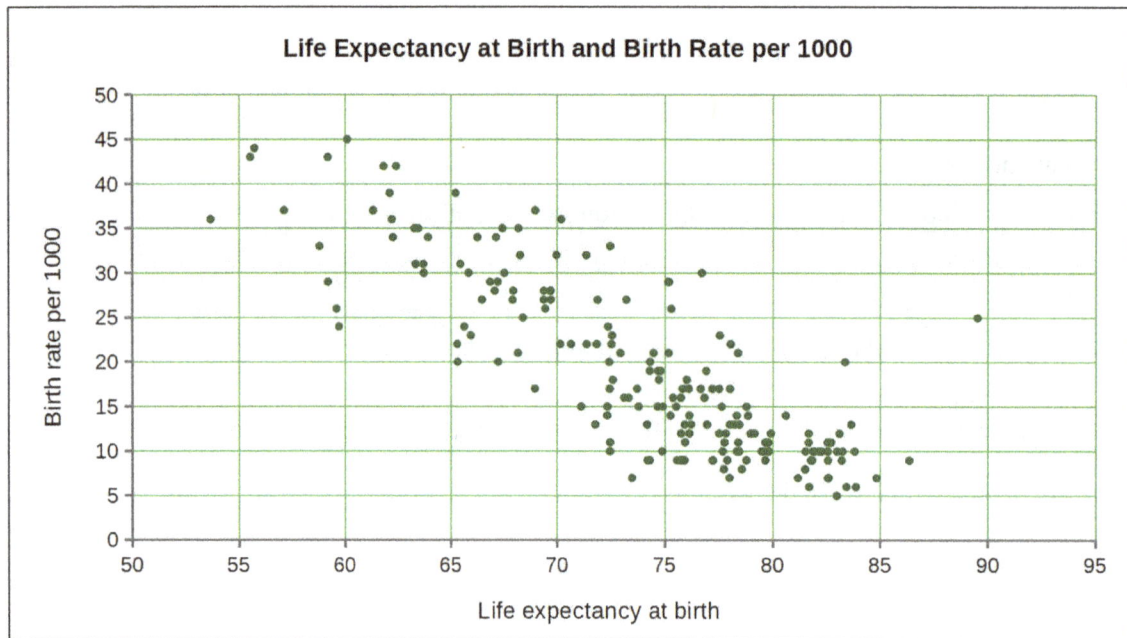

   **c.** Draw a line to represent the trend in the data.

   **d.** Find the (approximate) equation of your line.

   **e.** What does the slope of your equation signify in this context?

   **f.** Using your line, predict the life expectancy at birth in a country where the birth rate is 25 per 1000 population.

   **g.** Using your line, predict the birth rate per 1000 population in a country where the life expectancy at birth is 55 years.

   **h.** The dot at (89.5, 25) is Monaco. What is the difference between the predicted life expectancy at birth for a country with a birth rate of 25 per 1000, and the actual life expectancy of Monaco?

2. The scatter plot below has to do with the cost of airplane tickets and the distance covered by the flight. A spreadsheet program calculates the equation of the trendline to be $C = 0.069d + 76.9$ where $C$ is the cost (in dollars) and $d$ is the distance (in kilometres).

**Distance versus Airfare**

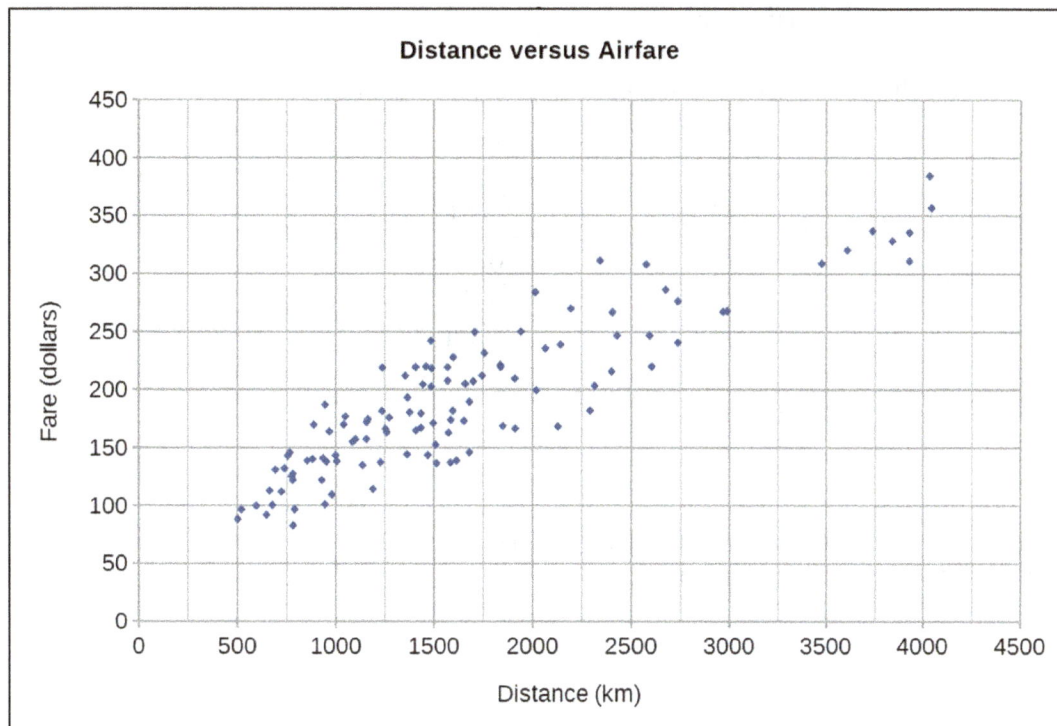

**a.** Sketch this line in the image.

**b.** Find four different points in the image for tickets costing slightly over 300 dollars. What are the distances covered in those flights?

**c.** What is the distance predicted by the trendline equation, if a ticket costs 310 dollars?

**d.** What is the ticket cost predicted by the trendline equation, if the flight distance is 1280 km?

**e.** For each 100-km increase in distance, how much would you expect the cost of the ticket to increase?

**f.** Based on the trend line, if you look at the airfares of $300 and $250, what is the associated difference in distance?

3. A multinational company asked its employees what was their favourite sport (to watch or to play). The results are categorized based on the location of the employees.

a. Calculate the totals. Then calculate the relative frequencies, to the nearest percent, based on the row totals, and write them in the table on the right.

|  | soccer | baseball | basketball | Total |
|---|---|---|---|---|
| Asia | 17 | 1 | 5 | |
| Europe | 85 | 7 | 60 | |
| Middle East | 40 | 3 | 10 | |
| North America | 11 | 45 | 26 | |
| Total | | | | |

|  | soccer | baseball | basketball | Total |
|---|---|---|---|---|
| Asia | | | | 100% |
| Europe | | | | |
| Middle East | | | | |
| North America | | | | |

b. Make at least three observations about the data. For one example, we can see that people in Middle East prefer soccer far above the other two sports.

c. Is there an association between a person's location and their preferred sport? Explain your reasoning.

d. If you randomly select a person who prefers soccer, how likely is it that the person is from Europe?

e. If you randomly select a person who prefers basketball, how likely is it that the person is from the Middle East?

f. If you randomly select a person from Asia, how likely is it that the person prefers basketball?

4. Which of the following numbers of 2, 9, or 24, would you put in the empty box so that...

a. there is no association between the variables?

b. girls are far more likely to play an instrument than boys?

c. boys are far more likely to play an instrument than girls?

|  | plays an instrument? | |
|---|---|---|
|  | yes | no |
| Boys | 12 | 18 |
| Girls | 6 | |

www.ingramcontent.com/pod-product-compliance
Lightning Source LLC
Chambersburg PA
CBHW080542220326

41599CB00032B/6331